UNDERSTANDING CYTOGENETICS

UNDERSTANDING CYTOGENETICS

By

Dr. Rajiv Tyagi

Department of Zoology

M.M. College

Modi Nagar (U.P.)

(India)

DISCOVERY PUBLISHING HOUSE PVT. LTD.

NEW DELHI-110 002

First Published-2009

ISBN 978-81-8356-462-5

Published by:

DISCOVERY PUBLISHING HOUSE PVT. LTD.

4831/24, Ansari Road, Prahlad Street,
Darya Ganj, New Delhi-110002 (India)
Phone: 23279245 • Fax: 91-11-23253475
E-mail: dphbooks@rediffmail.com
dphtemp@indiatimes.com
Website: www.discoverypublishinghouse.com

Printed at:

Sachin Printers
Delhi

Preface

The present title "Understanding Cytogenetics" has been written for those students interested in careers in diverse fields of biological sciences. It provides a structured approach to learning by covering all the important topics in a uniform, systematic format. The book has been comprehensively designed incorporating recent advances in this fast moving field. It also provides accessible information on cytogenetics in compact form for undergraduate students in biology and related life sciences. It is intelligible to the educated layman, though it deals with some complex ideas. It is an adequate text for all the requirements of students in this area. In addition, busy lecturers who require a quick reference compendium will find it useful, particularly for tuitional planning. Simple, yet hopefully clear figures and tables are provided throughout the book.

The over-riding goal of this book, and indeed of the whole *Understanding series,* is to present the essential information concerning cytogenetics in a compact, readily accessible form which leads itself to student learning and revision. The convergence of various approaches has generated a rich panorama of detail, the significance of which we are still attempting to unravel. The present text has been written as an introduction to this rapidly growing field.

To make the work more comprehensive and informative, the author has consulted many authoritative books, research journals, abstracts, monographs etc., so there can be no claim to originality except in the manner of treatment.

The author expresses his thanks to his friends and colleagues whose continue inspirations have initiated him to bring out this book.

The author expresses his gratitude to Mr. Wasan and staff of M/s Discovery Publishing House Pvt. Ltd. for their whole hearted cooperation in the publication of this book.

In the mean time, the author will remain sincerely responsible for any shortcomings of the book and be grateful to the readers for their suggestions and constructive criticism for the continuous betterment of the book. He takes this opportunity to appeal to the readers to send their suggestions straightaway to his Publisher.

Author

Contents

1

THE CELL

The cell is a unit of biological activity delimited by a semipermeable membrane and capable of self-reproduction in a suitable non-living medium. The body of all living organisms (bacteria, blue green algac, plants and animals) except viruses has cellular organization and may contain one or many cells. The organisms with only one cell in their body are called acellular organisms (e.g., bacteria, blue green algae, some algae, protozoan, etc.). The organisms having many cells in their body are called multicellular organisms (e.g., most plants and animals). Any cellular organism may contain only one type of cell Prokaryotic cells, Eukaryotic cells.

It was *Robert Hooke* who first of all in 1665 observed under a microscope certain honey comb like structures in cork, and he applied the term 'cell' (L., *cella;* small room) for the same. Previously it was believed that there is only one component of the cell which is cell wall. But in 1831, one more peculiar structure was observed by *Robert Brown.* He gave the name 'nucleus'. Later on the 1835 the word 'Sarcode' was proposed for the jelly like material present inside the cell by *Dujardin.* In 1840 *Purkinje* replaced the word sarcode by Protoplasm which is now used universally.

PROKARYOTIC CELLS

The prokaryotic (Gr., *pro*-primitive, *Karyon*-nucleus) cells are the most primitive cells from morphological point of view. They occur in bacteria and blue green algae. Prokaryotes are small, single cells organism, usually less than a micrometer (abbreviated μm; 1000 μm = 1 millimeter, abbreviated mm) are generally not longer than 3 μm. Their structural organization is simple than that of the eukaryotes.

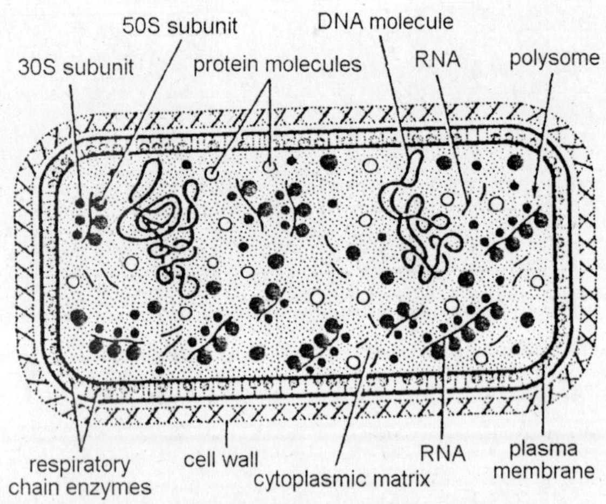

Fig. 1.1. A prokaryotic cell of Escherichia coli.

They do not have a distinct nucleus, and they are lacking in many membranous structures found in the more complex cell types. The hereditary information is contained in a single circular structure that lacks many of the molecular components of the eukaryotic hereditary structure called the chromosome. In addition, the prokaryotes seem to lack any kind of special apparatus for cell division that is why they divide amitotically or by fission.

The prokaryotes are characterized by:

(i) the absence of a membrane around the nuclear material.

(ii) the absence of clearly defined membrane-limited organelles like mitochondria, chloroplast, Golgi complex and lysosomes.

(iii) the genetic material is located on a single chromosome which consists of a circular double strand of DNA.

(iv) the basic protein—histones, which are one of the most important constituents of chromosomes of eukaryotic cells, are absent in prokaryotic chromosomes.

(v) the absence of nucleolus and mitotic apparatus.

(vi) the cell wall is non-cellulosic, being formed of carbohydrates and amino acids.

(vii) the plasma membrane which lies below the cells wall is produced into the cytoplasm and acts as the mitochondrial membrane carrying respiratory enzymes, and

(viii) the cytoplasm neither exhibits streaming nor the amoeboid movement.

The prokaryotes include bacteria, blue green algae, spirochaete, *Mycoplasma* or pleuropneumonia like organism (PPLO).

Bacterial Cell

The bacteria are microscopic, cellular, achlorophillous, asexually reproducing prokaryotes which occur in fresh and salt water, in soil and in plants and animal. They lead either a saprophytic or parasitic mode of existence. The saprophytic species of bacteria are of great economic significance for man. Some parasitic species of bacteria are pathogenic (disease producing) to plants, animals and man.

Bacteria are very small having an average diameter of 1.25μ.

The smallest bacterium is *Dialister pneumosintes* (0.15 to 0.3μ) and largest bacterium is *Spirillum volutans* (13-15μ) in length.

One of the best studied bacteria is *Escherichia coli*. It is about 2 μm long and 1μm wide and contains several thousand different kinds of molecules. Structurally, *E. coli* is rod like and has a typical cell wall and underlying cell membrane. The membrane contains the special regions called *mesosomes,* which apparently have a role in cellular respiration and cell division. There may be one or several "nucleoid" areas, which contain a single, circular molecule of hereditary information (DNA).

The rest of the bacterial cytoplasm is full of structures called ribosomes, which differ slightly from the ribosomes of eukaryotic cells. These structures are the sites of protein synthesis. In addition, recent studies have revealed the presence of small circular DNA molecules in the cytoplasm called *plasmids*. These structures are very important in genetic engineering.

Mycoplasma

The *Mycoplasma* (previously called PPLO, pleuropneumonia like organisms) are small (0.2 to 0.3 μm) and more simply organized than bacteria. The hereditary material (DNA) is contained in a non-membrane-bound, nucleus like region, and it may exist either as strands or as a circular molecule as in the bacteria. However, there is only one tenth the amount of genetic material in Mycoplasma as there is in bacteria. Similarly, the Mycoplasma contains 1/50 to 1/100 the number of bacterial ribosomes per cell. A delimiting cell membrane exists, but no cell walls have been detected. A variety of other cytoplasmic inclusions, such as vacuoles and granules, have been detected, but their function is not known. As in other prokaryotes, there are no intracellular membranous structures.

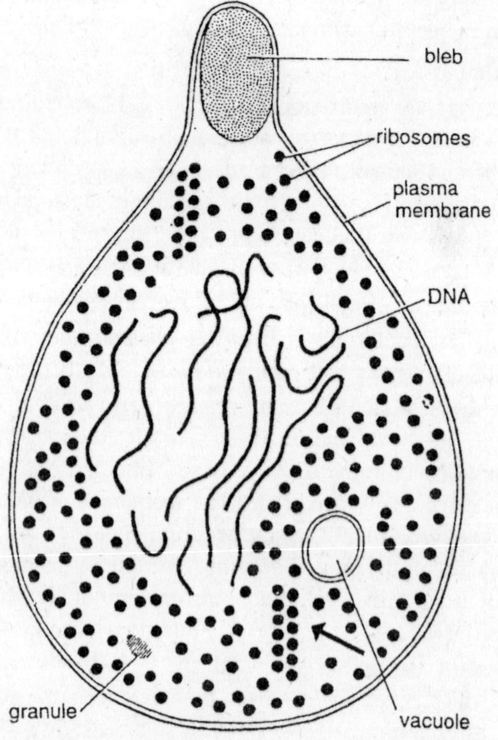

Fig. 1.2. A schematic diagram of typical PPLO cell.

Mycoplasma has been known to cause certain respiratory diseases in animals and man. They also occur as routine contaminants in eukaryote tissue culture and currently are of much concern to researchers who use these culture procedures in their studies.

Blue-Green Algae

The blue-green algae are the third major group of prokaryotic organisms. The cellular organization is more comples than that of the other prokaryotes, primarily because of the existence of photosynthetic machinery. The blue-green algae contain chlorophyll as well as other more unusual pigments called *phycobilins.* The chlorophylls are bound to membranes in flattened rows called lamellae, which give the algal cytoplasm a rather organized and structured appearance. The phycobilins appear to be located in granular structures called *cyanosomes,* which apparently are attached to the outer lamellar membrane surface. Like the other prokaryotes, the blue-green algae do not have a membrane-bound nucleus or other membrane-delimited organelles. The genetic

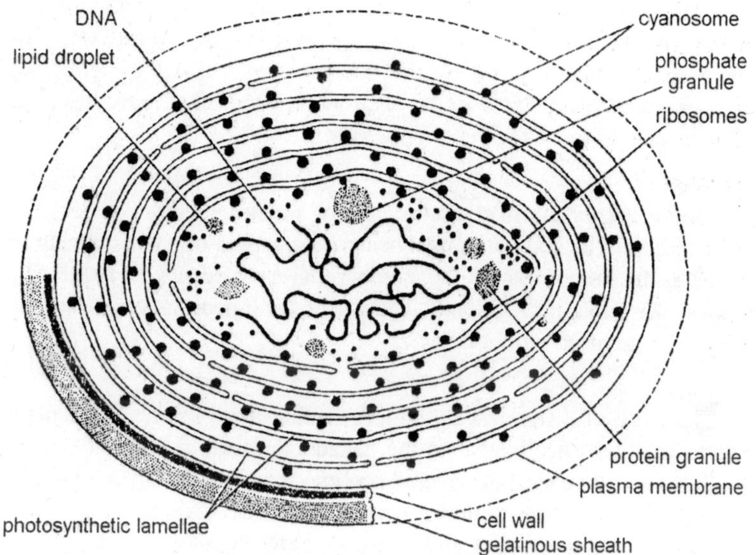

Fig. 1.3. A prokaryotic cell of cyanobacteria (electron microscope view).

material exists as free strands of DNA, and the ribosomes are free in the cytoplasm as in the bacteria and *Mycoplasma.* There appears to be other cytoplasmic granules of a protein and phosphate nature, and the entire cell is surrounded by a cell wall with an underlying plasma membrane.

Eukaryotic Cells

Eukaryotic cells may be acellular organisms, such as protozoans and acellular algae, or they may be cells that make up the tissues and organs of multicellular organisms. Though the eukaryotic cells have different shape, size, and physiology but all the cells are typically composed of plasma membrane, cytoplasm and its organelles, viz., mitochondria, endoplasmic reticulum, ribosomes, Golgi complex, etc., and a true nucleus.

Morphology of Eukaryotic Cell

Shape

Cell shapes are almost as numerous as cell types: there is not typical shape. The cells of certain unicellular forms, such as *Amoeba, Diatoms, Acetabularia* and bacteria exhibit a number of shapes. But generally the cells are rounded or spherical. Besides this the cells like oval, cuboidal, cylindrical, flat, discoidal, polygonal have also been observed.

The shape of cells depends mainly on functional adaptations and partly on the surface tension and viscosity of the protoplasm, the mechanical action exerted by the adjoining cells and the rigidity of the cell membrane. Some cells, such as *Amoeba* and leucocytes, can change their body shape very frequently.

Number

Some organisms like protozoans are single celled and others are multicellular. The body of human being is composed of about 26 trillions of cells. In human blood the number of erythrocytes is about five million per cubic ml of blood. About 10 billion neurons constitute the nervous system in human being.

Size

The size of different cells ranges within broad limits. Some plant and animal cells are visible to the naked eye, such as eggs of certain birds have a diameter of several centimeters. But the great majority of cells are visible only under microscope. The smallest living cells are found among bacteria and viruses where the size ranges 0.1μ to 1μ (1μ = 0.01 mm). Organisms like Pleuropneumonia-like organisms so called "elementary bodies" have been observed with diameters of 100 mμ. These appear to be a resting from of the bacteria which may grow into bodies of 250 mμ in diameter during the active metabolism of the organism. It would seem that 200-250 mμ in diameter is the lower limit for the size of inactive, living cells. This lower limit may be set by the minimum number and size of the component necessary for independent cellular existence. Usually the vast majority of cells lie in the range of .5 to 20μ in diameter. The length of acellular diatoms is about 100μ. And the size of *Amoeba proteus* is 1 m.m. (1000μ) in length. The size of human R.B.C. is 7-8μ in diameter.

The long nerve is also an example of the extremes in cell sizes, the range of which is quite remarkable in the living world.

Factors Restricting Cell Size

In general, the size of the cell is correlated with its function. In no way it is associated with the size of the organism. For instance, a microscopic cell of bread mould can be several metres long, whereas cells in the body of whale or elephant may be as small as 10μ. Two major factors seem to restrict cell size.

Nucelocytoplasmic ratio

The nucleus controls activities of the cell by protein synthesis through RNA synthesis. In a large cell cytoplasm requires more proteins and consequently more RNA. The DNA content of a cell is constant, which can control the activities of the cell upto a specific size, beyond which proper control is not possible. For this reason a cell after attaining a specific size stops growing.

Surface area

As shown in fig, the small cells have more surface area per unit volume and with an increase in size, this ratio between surface area and volume gradually decreases. However, a large cell naturally requires more nutrients and oxygen for its metabolism, which diffuse into the cell through its surface. It means, the surface area of a cell affects its ability to exchange materials with its environment and puts a check on cell volume. For this reason, the cells that are metabolically more active are small in size.

Similarly, large cells face problem in the diffusion of excretory products from cell cytoplasm to the exterior.

Structure of Fixed Cells

Examination of the living cell is limited to light microscopy and is based mainly on the differences in refractive index of the different cell components. Sometimes the use of stains that act on the living organism (vital staining) facilitates observation of the living cell. However, more important in the morphologic study of the cell are methods of fixation, by which cell death results in such a way that physiologic structure and chemical composition are preserved as much as possible.

The complexity of the structural organization in a cell of the higher plants and animals is most impressive. Although there are great differences between the primitive forms of life, and the higher plant and animal cells, the similarities between primitive and advanced cells are also notable. Eukaryotic cells are characterized by a true nucleus with a nuclear membrane or envelope which divides the cell into two main compartments: nucleus and cytoplasm. The cytoplasm is turn is limited by the plasma membrane. In a plant cell the plasma membrane is covered and protected on the outside by a thicker cell wall through which there are tunnels, the *plasmodesmata,* by which the cell inter-communicates with neighbouring cells via fine cell process. In animal cells, parts of the plasma membrane are covered by a thin layer of material, which is generally described as the extraneous coat of the plasma membrane. The co-called *basement membranes* correspond to this extraneous coat.

In every stained cell usually one nucleus is present. It is central organelle of cell and is the main site of hereditary material. The nucleus is surrounded by a nuclear membrane but in bacteria and blue green algae it is without nuclear membrane. Within the nucleus a single nucleolus and network of chromatin is present. These two materials remain suspended in the nucleoplasm.

A small cytoplasmic organoid 'centriole' is present in the majority of animal cells and in some cells of the lower plants. The position is generally fixed for each type of cell. Generally the cell center or centriole occupies an axial position behind the nucleus. In some cells, it has the tendency to occupy the geometrical center i.e., in the leucocytes it occupies this position within the nucleus which is horseshoe shaped. The morphology of centriole changes with the functional state of the cell.

It is cytoplasm of the cell, there are several organelles out of them, only mitochondria and Golgi bodies are visible under light microscope. Mitochondria or chondriosomes are scattered in the cytoplasm of all kinds of cells. Their shape in the fixed cells depends upon the fixation and handling of the tissues before preparation they may appear as long rods, short rods or as small granules. Golgi material appears in its very prominent form in nervous and secretory cells in animals. Plastids are cytoplasmic organelles of plant cells may or may not contain pigments. In green plants chlorophyll is present in these organelles and thus termed as chloroplasts.

Electron Microscopic Structure of a Typical Cell

Under ordinary light microscope only few cell organelles like mitochondria, Golgi complex, chloroplasts and nucleus are visible. However, under electron microscope, several other cytoplasmic organelles such as endoplasmic reticulum, ribosomes, lysosomes, nuclear membrane, plasma membrane, etc. are seen.

Some important cellular components are being described here under the following heads:

Plasma membrane

The outermost boundary of the animal cell is called the 'plasma membrane'. It is invisible under a light microscope but under electron microscope, it appears to be composed of two dense layers which are called the outer dense layer and inner dense layer. Both these layers are about 20Å in thickness, and are separated from each other by a less dense area of about 35Å in thickness. By this way the total thickness

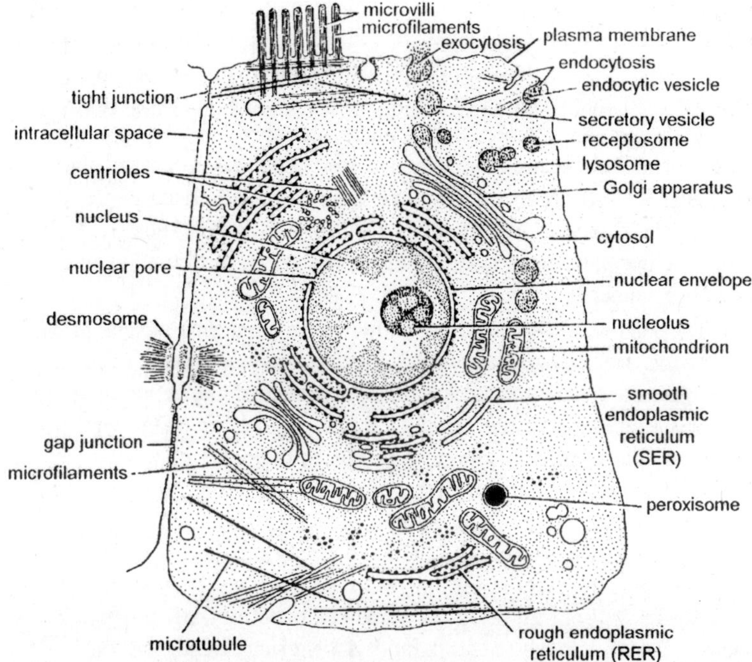

Fig. 1.4. Ultrastructure of a typical animal cell as seen in the electron microscope.

of plasma membrane is about 75Å. Before the advance of electron microscope people used to think of the plasma membrane has been stretched tightly over the cell. But now we know that a single plasma membrane can have as many as 3000 microvilli. Another peculiarity of the structure of the membrane is that, it remains connected with the endoplasmic reticulum.

Plasma membrane is lipo-proteinaceus in composition.

Functions

(i) Gives an identity to the cell.

(ii) Protects the cell organelles.

(iii) Helps in digestion through phagocytosis and pinocytosis.

(iv) Being permeable to water, water can pass through it.

(v) Help in the transport of ions.

(vi) Helps in the formation of cytoplasmic organelles.

Mitochondria

The term mitochondria was, for the first time used by *Benda* (1902) to designate the filamentous and granules of cell cytoplasm.

Most of the knowledge related to mitochondrial fine structure is due to the studies of *Palade* (1952, 55, 56) and *Sjostrand* (1955). Due to these studies, now mitochondria are defined as the double walled structures of cell organelles within which the membranous structures called *cristae* are present. The outer chamber of mitochondrion is made up of two dense layers and a less dense layer having the total thickness of 140 Å to 180Å while the outer and inner dense layers vary in their measurements from 35Å to 60Å in thickness depending upon the physiological state of the cell. The less dense layer is 40-47Å in thickness.

The matrix of the mitochondria may have some fine opaque granules of various size. *David E. Green* (1964) observed in a very highly magnified electron microscope that the inside of the inner membrane and the outside of outer membrane are covered with very small particles. They measure about 90-100Å in diameter. The particles on the outer membrane differ from those on the inner membrane. The former generally appear as simple spheres packed closely together which give to the surface of the mitochondrion, a rough or pimpled appearance. The particle on the inside of the inner membrane has the more complex configuration. In some cells, but not all, this particle is seen to have a base, a stalk and a spherical head. Mitochondria represent the power generating centres of the cell. They can bring about oxidation of the intermediates of carbohydrates, fat and amino acid metabolism in the presence of O_2 releasing energy. The released energy is trapped in the form of adenosine triphosphate (ATP).

Golgi Complex

Golgi comples have probably been first observed earlier in 1867 by *La. Vallete St. George*, but any how credit goes to *Camillo Golgi* (1898) who described it as internal reticulate apparatus' and after it, was named *Golgi apparatus*. It is variously called Golgi element, Golgi complex, Dalton's complex etc. Each Golgi complex is composed of many lamella, tubules, vesicles and vacuoles. The membranes of Golgi complex are of lipoproteins and these are supposed to be originated from the membranes of endoplasmic reticulum. The function of the Golgi complex is storage of proteins and enzymes which are synthesized by ribosomes and transported by endoplasmic reticulum to them. Further the Golgi complex has most important secretory function. It secretes many secretory granules and lysosomes. In plant cells the Golgi complex is known as dictyosome and secretes necessary materials for cell wall formation during cell division.

Lysosomes

The lysosomes (Gr. *lyso;* digestive+*some;* body) are also found scattered in the cytoplasm. By gently centrifuging lysosomes can be separated from other cell organelles. They are usually spherical but may be irregular in shape, covered with a membrane of lipoprotein. Inside structure is always variable. Lysosomes contain hydrolysing enzymes, such as acid phosphatase, acid ribonuclease, acid deoxyribonuclease and cathespins. But they do not have oxidative enzymes. This property distinguishes them from the mitochondria. Lysosomes are polymorphic in nature and are commonly known as suicide bags and are responsible for postmortem changes in cells.

Endoplasmic Reticulum

The term "Microsome" was introduced to the modern cytology by *Claude* (1943) to represent one of the submicroscopic cellular components isolated by centrifugation. These are now known to be the broken parts of the endoplasmic reticulum after the studied of *Kollman* (1953). *Porter* (1945) was the 1st man to describe their electron microscopic structure in cultured cells. He described them as lace-like reticulum. These are membrane bounded sacs in the form of double membrane or cisternae. The term cisternae was introduced by *Sjostrand* (1953) to describe long or elongated rod like parts of the endoplasmic reticulum measuring 50-200Å. *Wiess* (1953), described the vesicles of the endoplasmic reticulum which usually has a diameter of 25-500Å while *Bradfield* (1913) came across of another type of endoplasmic reticulum termed them as tubules, they measures from 50 to 100Å in diameter.

On the basis of presence or absence of ribosomes or RNP particles the endoplasmic reticulum (E.R) is distinguished in two varieties. The *rough walled endoplasmic reticulum,* and *smooth walled endoplasmic reticulum.*

Rough Walled Endoplasmic Reticulum

Rough walled endoplasmic reticulum (R.E.R.) is that variety which bears the ribosomes at the external surface of the cisternae. The distribution of the ribosomes can be circular, spiral, or rosette type. These particles may be induced to leave the surface of the cisternae without damaging the structure of the cisternae or the capacity to associate again with the particles. Usually the rough walled endoplasmic reticulum is richly distributed in those cells which are engaged in the synthesis of the protein.

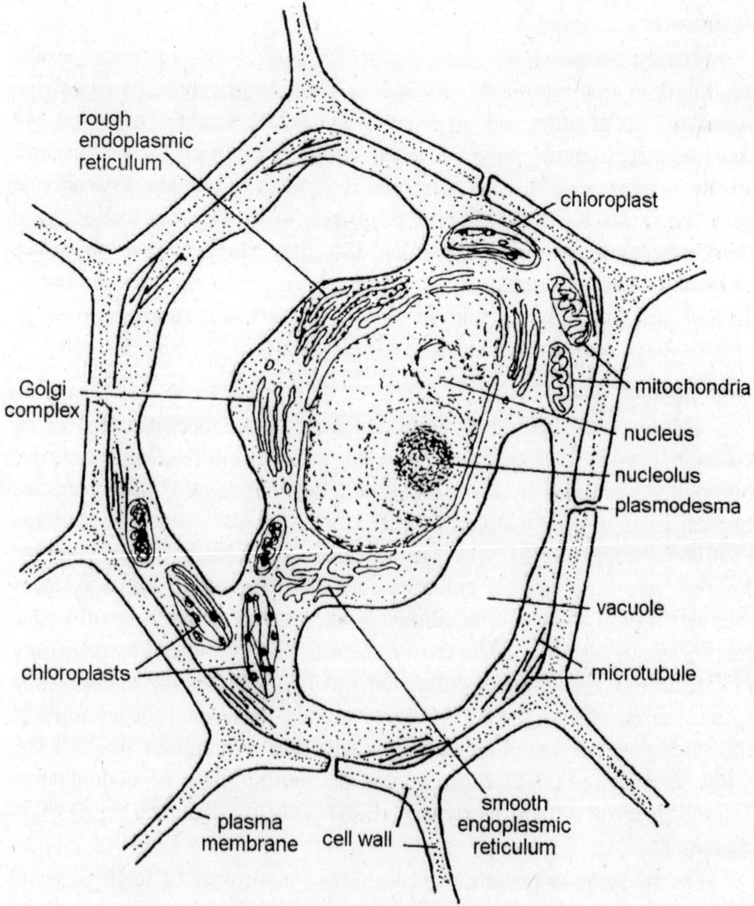

Fig. 1.5. Electron microscopic structure of a plant cell.

Smooth Walled Endoplasmic Reticulum

The other large division of the system owes its identity in parts of the absence of the particles and therefore commonly referred to as the smooth surfaced a granular form. Like the R.E.R., the S.E.R. shows a characteristic morphology which is tubular rather than cisternae. The tubules have the diameter nearly of 50-100 mμ. As studied by *Fawcett* (1960) S.E.R. is richly distributed in those cells which are engaged in the synthesis of steroids.

Continuity between the S.E.R. of a particular cell type possess a characteristic pattern of structure. Thus it is typical for rat liver cells to show groups of eight or ten slender profiles in parallel array.

Plastids

Plastids are present only in plant cells (not in animal cells). These are double-membrane structures relatively larger than the mitochondria. Some of the plastids are colourless and store starch (Leucoplasts). Other contain various pigments and impart different colours (red, brown, yellow, etc.) to different parts of the plant. These are collectively called chromoplasts. Those containing green pigment—cholorophyll, are called choloroplasts. These contain several enzymes in their matrix necessary for photosynthesis. The chloroplasts capture light energy of the sun and convert it into the chemical energy—ATP, which is used in the biosynthesis of glucose from CO_2 and H_2O.

Ribosomes

Many minute, spherical structures known as *ribosomes.* They remain attached with the membrane of endoplasmic reticulum and forms granular or rough type of E.R. The ribosomes are originated in the nucleolus and consist of mainly the ribonucleic acids (RNA) and proteins. Each ribosome is composed of two structural units, small subunit known as 40S subunit and a larger subunit known as 60S subunit. The ribosomes remain attached with the membranes of endoplasmic reticulum by the 60S subunits. The 40S subunits occur on the larger sub unit and form a cap-like structure. The ribosomes consists of 3 types of RNA's known as ribosomal RNAs or rRNA, viz., 5S, 18S and 28S rRNAs. The 28S and 5S rRNAs occur in large (60S) subunit, while 18S rRNA occurs in the smaller ribosomal subunit. The ribosomes also occur freely in the cytoplasm. They are the sites of protein synthesis.

Peroxisomes

These are also bounded by a single membrane. These contain enzymes for the breakdown of hydrogen peroxide to form water and oxygen. These are protective organelles of the cell.

Cilia and Flagella

The cells of many unicellular organisms and ciliated epithelium of multicellular organisms consists of some hair-like cytoplasmic projections outside the surface of the cell. These are known as cilia or flagella. They help in locomotion of the cell. The cilia and flagella consists of nine outer fibrils around the two large central fibrils. Each outer fibrils consists of two microtubules. The two microtubules of each peripheral pair or *doublet* remain morphologically distinct. The central fibrils are separate but remain enveloped in a common *central sheath.* Cilia and flagella are envolved in locomotion, feeding, cleansing etc.

Vacuoles

These structures are fluid-filled "bubbles" found in the cytoplasm. Vacuoles are bounded by a membrane, called *tonoplast,* which is structurally similar to, if not identical with, a cell-membrane. Inside the vacuoles, food materials or wastes are found.

Vacuoles are consipicuous in plant-cells but they are found less in animal cells. Vacuoles are absent in actively dividing cells. In a young plant-cell, there are many vacuoles but, when a cell matures, they unite to form a single, large, central vacuole.

Guilliermond (1941) was of the opinion that vacuoles are colloidal with a strong imbibing capacity. Vacuoles arise *de novo* i.e., spontaneously formed from pre-existing vacuoles.

In many protozoan, vacuoles are formed at the bottom of the gullet. In these vacuoles, food-materials, such as bacteria, are collected. Some vacuoles are formed by cell-drinking. This type of vacuoles are called *pinosomes.* Food-vacuoles formed by evagination of cell-membrane surrounding food-particles with some fluid, are called *phagosomes.* In animals cells, the vacuoles are formed by a lipoprotein membrane that helps in storage and transport of materials and in the maintenance of internal pressure of the cell.

Vacuoles are not a living part of the protoplasm but are a storing place of reserve materials and waste products.

Crystals and oil droplets

In certain plant cells the food or waste material is found deposited in the form of crystals. The oil droplets are round in both plant and animal cells, which appear as tiny glistening white spheres. These serve as a reserve supply of energy rich fuel.

Nucleus

It is the most vital part of cell which is invariably present in all the cells of higher animals and plants. Normally one nucleus is present in a cell but more than one may be present in certain cells.

Structurally it is enveloped by a porous *nuclear membrane* that separates it from the surrounding cytoplasm. It has the characteristic permeability. During cell division, it breaks up and soon after the mitosis it is reconstituted. The inside of nucleus is filled with *nuclear sap* or *karyolymph.* In nuclear sap remain embedded other nuclear constituents. The *chromatin* is the Feulgen positive material in the interphase nucleus which during division becomes the definite chromosomes. In short it represents the chromosomes substance. The

Table 1.1. The relative volumes occupied by the major intracellular compartments in a typical liver cell (hepatocyte)

Intracellular Compartment	*Percent of Total Cell volume*	*Approximate Number per Cell*
Cytosol	54	1
Mitochondria	22	1700
Rough endoplasmic reticulum cisternae	9	1
Smooth endoplasmic reticulum cisternae plus Golgi cisternae	6	
Nucleus	6	1
Peroxisomes	1	400
Lysosomes	1	300

sex-chromatin are the specific heterochromatic bodies, found generally near the periphery of the nucleus. In nucleus usually one sex chromatin body is found but more than one can also be observed. *Nucleolus* is relatively large, generally spherical ball-like body. It has a high concentration of RNA and thus is an important structure involved in protein synthesis. Its number in a nucleus vary in different cells and depend on either the species or the sets of chromosomes.

The nucleus co-ordinates many cell activities. DNA molecules are the source of coded instructions, which pass to RNA molecule which then act as messenger, moving out of the nucleus to one of the various structures of the cytoplasm. It also controls the cell reproduction. Here too, DNA is the source of coded instructions.

The Cytosol, Hyaloplasm or Ground Substance

The material that remains suspended under conditions sufficient to sediment even the ribosomes is referred to by biochemists as the *soluble fraction* or *cytosol.* This fraction usually contains some material lost from the organelles in the course of homogenization and centrifugation. However, it also contains other molecules, including some enzymes, that are not known to be part of the intracellular structures indentifiable in the microscope. Such molecules are usually thought to derive from the *hyaloplasm* (also called ground substance, cell matrix, or cell sap), which was originally defined by microscopists as the apparently structureless medium that occupies the space between the visible organelles. As will be evident in subsequent chapters, with the improved techniques less and less of the cell appears structureless. Most investigators continue to believe that there does exist in the cytoplasm a "soluble phase" – a solution containing inorganic ions and small

Table 1.2. Comparison of prokaryotic and eukaryotic organisms

	Prokaryotes	*Eukaryotes*
Organisms	bacteria and cyanobacteria	protists, fungi, plants, and animals
Cell Size	generally 1 to 10 μm in linear dimension	generally 10 to 100 μm in linear dimension
Metabolism	anaerobic or aerobic	aerobic
Organelles	few or none	nucleus, mitochondria, chloroplasts, endoplasmic reticulum, etc.
DNA	circular DNA in cytoplasm	very long DNA containing many non-coding regions; organized into chromosomes and bounded by nuclear envelope
RNA and protein	RNA and protein in same compartment	RNA synthesized and processed in nucleus; proteins synthesized in cytoplasm.
Cytoplasm	no cytoskeleton, cytoplasmic streaming, endocytosis, or exocytosis	cytoskeleton composed of protein filaments; cytoplasmic streaming; endocytosis and exocytosis
Cell division	by binary fission	by mitosis (or meiosis)
Cellular organization	mainly acellular	mainly multicellular, with differentiation of cells.

molecules, materials in transit from one cell region to another and dissolved or suspended enzymes that function as individual molecules or perhaps or parts of small groups of molecules rather than as components of large, microscopically recognizable organelles. But there is a lively debate going on over proposals that many components once though to be suspended in this solution are actually bound, perhaps loosely, in structured arrangements that do not survive the disruption involved in cell fractionation.

Table 1.3. Comparison of plant and animal cells

Plant Cell	*Animal Cell*
1. Generally larger than the animal cell.	1. Usually smaller than the plant cell.
2. Plant cells are surrounded by a rigid cellulose wall.	2. Cellulose wall is not found in animal cell.
3. Majority of the mature plant	3. Vacuoles in animal cells are small.

cells have a large central sap vacuole.	
4. Plastids, especially chloroplasts are found usually in cells of green plants.	4. Chloroplast is not found in animal cell except in some Protozoa.
5. In the cell division, the division of cytoplasm of plant cells take by formation of a partition called place cell plate.	5. Animal cells divide by constri-ction during cell division.

Table 1.4. Functions of cellular organelles

Cell Organelles	*Functions*
Cell or Plasma membrane	Serves as a differentially permeable mem-brane through which extra-cellular sub-stances may be selectively sampled and cell products may be liberated. Gives external form. Provides mechanical support and thus protects the cell.
Cell Wall (in plants)	Thick cellulose wall around the cell mem-brane gives strength and rigidity of the cell.
Cytoplasm	Provides greatly expanded surface. Endoplasmic recticulum-area for various biochemical reactions which normally take place at or across membrane surfaces.
Ribosomes	Sites of protein synthesis and known as "Protein factories on the cell"
Mitochondria	Oxidation of food substances, changes potential energy into a form of energy to be used by cell. Energy production (Krebs' cycle, electron transport chain, beta oxidation of fatty acids etc.). Power house of cell.
Centrioles (usually not seen in plant cells)	Form poles for the cell division. Capable of replication usually.
Golgi body or apparatus	Produce cellular secretions known as dictyosomes in plants.
Lysosome (in animals)	Produce intracellular digestive enzymes which help in disposal of bacteria and other foreign bodies. If ruptured, may cause cell destruction.
Plastids (in plants)	Serve for storage of starch, pigments and other cellular products. Photosynthesis takes place in chloroplasts.
Vacuoles	Store excess water, waste products, soluble pigments etc.

Hyaloplasm	Contains enzymes for glycolysis and structural materials, viz., sugars, amino acids, water, vitamins, nucleotides etc.
Cilia and flagella	Serve to move the animals or to move the material.
Nucleus	Provides selective continuity.
Nuclear membrane	between nuclear and cytoplasmic materials.
Nucleoplasm	Contains materials, for building DNA and messenger molecules which serve as intermediates between nucleus and cytoplasm.
Nucleolus	May synthesize ribosomes. Disappears during cellular replication. Function uncertain.
Chromosomes	Bearers of hereditary instructions. Regulation of cellular processes (evident only during nuclear division). Control the nucleus and metabolism.

2

Cell Membrane, Structure and Function

Consider the challenge presented to a unicellular organisms drifting about in a tide pool, stranded by the receding tide. It inhabits an environment whose conditions change quickly and often. As the tide recedes, the sun warms the pool. As the day wears on, some of the water evaporates, boosting the concentration of dissolved salts. A sudden rain or the returning tide dilutes the water again. When night falls, the water temperature falls sharply. Just surviving requires a protective barrier between the inside the outside of the cell.

Merely withstanding environmental changes is not enough, however. The cell must absorb sunlight for photosynthesis or engulf other organisms to obtain the nutrients it needs to make new cellular components. To maintain a consistent internal environment in which its components function best, the cell must shuttle salt, water, and other molecules in and out. It must excrete waste products. It must detect chemical clues to the location of food or to healthy and unhealthy environmental conditions. Finally, if it reproduces sexually, it must communicate with at least one other cell of its own species sometimes during its life.

A lipid membrane only tow molecules thick stands between the delicate inner working of the cell and its hostile environment. The membrane also selectively permits the absorption and release of chemicals, sometimes using cellular energy. It allows chemical reactions to occur between one molecule outside the cell and another inside the cell. And it may deform to engulf food or enable the cell to move.

All these processes rely on a variety of proteins embedded in the lipid membrane or attached to its surfaces.

Some single-celled organisms, such as the amoeba, survive with no barrier between life and the environment but a cell membrane. Cells of plants, bacteria, and many protists, such as diatoms, have walls outside the cell membrane that lend mechanical support. Cells of multicellular animals live surrounded by other cells possess very similar cell membranes and carry on many of the same processes across that membrane.

In this chapter, we explore the cell membrane structures that make defense, exchange of materials with the environment, and communication between cells possible.

Cells Walls

The outer surface of the cells of bacteria, plants, fungi, and some protists are covered with stiff, nonliving coating called *cell walls*. Plant cell walls are composed of cellulose and other polysaccharides, while fungal cell walls are made of the modified polysaccharide chitin. Bacterial cell walls have a chitinlike framework to which short chains of amino acids and other molecules are attached.

Cell walls are produced by the cells they surround. In plants, membranous vesicles filled with sticky polysaccharides such as protein (the ingredient that congeals grape juice into jelly) line up across the

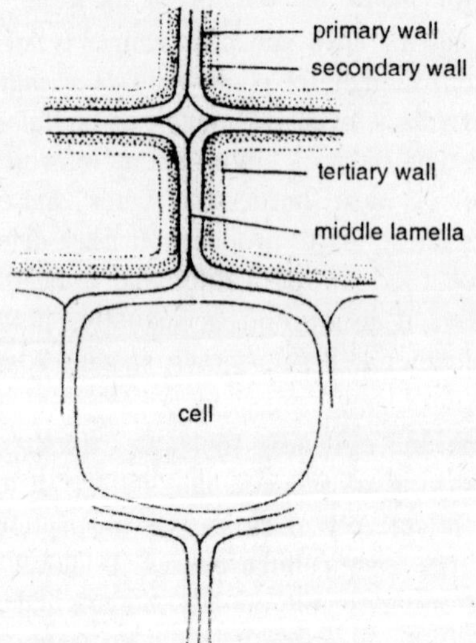

Fig. 2.1. Structure of plant cell wall.

middle of a dividing cell. The vesicles fuse together, forming the new plasma membranes that separate the two daughter cells. The pectins formerly contained in the vesicles glue the daughter cells together, forming the *middle lamella*. The two cells then secrete cellulose through their plasma membranes, underneath the middle lamella, forming the *primary cell wall*. Many plant cells later secrete more cellulose and other polysaccharides beneath the primary wall to form a thick *secondary cell wall*. In some plant cells, the secondary wall may become thicker than the entire rest of the cell.

Cells walls support and protect otherwise fragile cells. For example, cell walls allow plants and mushrooms to resist the forces of gravity and wind and to stand erect on land. Tree trunks are the ultimate proof of the strength of cell walls, being composed almost entirely of cellulose and other materials laid down over the years and capable of supporting impressive load.

Although strong, cell walls are usually porous, permitting easy passage of small molecules such as minerals, water, oxygen, carbon dioxide, amino acids, and sugars (otherwise, the cell within would soon die). However, the structure that really governs the interactions between a cell and its external environment is the plasma membrane.

Plasma Membrane

The *plasma membrane* of a cell can be thought of as a gatekeeper, allowing only specific substances in or out and passing messages from the external environment to the cell's interior. As gatekeeper, the plasma membrane must perform several specific functions:

1. Isolate the cell cytoplasm from the external environment.
2. Regulate the exchange of essential substances between the cytoplasm and the external environment.
3. Communicate with others cells.
4. Identify the cell as belonging to a particular species and a particular individual member of that species. In multicellular organisms, specific cell types often also bear unique molecular labels on their cell surfaces.

These are formidable tasks for a structure so thin that 10,000 stacked atop one another would scarcely equal the thickness of this page. The key to membrane function lies in membrane structure. Membranes are not simply homogeneous sheets: They are complex, heterogeneous structure, with different parts, indeed even individual molecules, performing very distinct functions.

All the membranes of a cell have a similar structure. Therefore, although this chapter focuses on the plasma membrane, the information presented here applies to other cellular membranes as well.

Fluid Mosaic Model Describes the Structure of the Plasma Membrane

According to the *fluid mosaic model* of cellular membranes, developed by cell biologists S.J. Singer and G.L. Nicolson in 1972, a membrane, when viewed from above looks something like a lumpy, constantly shifting mosaic of tiles. A bilayer of phospholipids forms a viscous, fluid "grout" for the mosaic, while an assortment of proteins are the "tiles," often sliding sluggishly about within the phospholipid bilayer. Thus, the overall image slowly changes over time, though the components remain relatively constant. As strange as this mode may seem, it captures something of the dynamic quality of real membranes. With this model in mind, let's look more closely at the structure of membranes.

Fluid Portion of Membranes is Produced by a Double Layer of Phospholipids: The Phospholipid Bilayer

A phospholipid consist of two very different parts, a polar hydrophilic head and a pair of nonpolar hydrophobic tails.

All living cells are surrounded by water, whether it is the pond in which an Amoeba spend its life or the extracellular fluid that bathes the cells of animals. The cell cytoplasm is also mostly water. Plasma membranes therefore separate a watery cytoplasm from a watery external environment. Under these conditions, phospholipids spontaneously arrange themselves in a double layer called a *phospholipid bilayer*.

Hydrogen bonds can form between water and the phospholipid heads, so the heads face the cytoplasm or the extracellular fluid. Hydrophobic interactions cause the phospholipid tails to hide inside the bilayer. Because individual phospholipids molecules are not bonded to one another, this double layer is quite fluid, and individual molecules move about easily.

Most biological molecules, such as salts, amino acids, and sugars are polar and water soluble. They therefore cannot pass easily through the nonpolar, hydrophobic fatty acids tails of the phospholipid bilayer. Because most substances that contact a cell are water soluble, the phospholipid bilayer acts as a barrier to the entrance of these molecules. It is largely responsible for the first of the four membrane functions that we listed earlier—isolating the cell cytoplasm from the external environment.

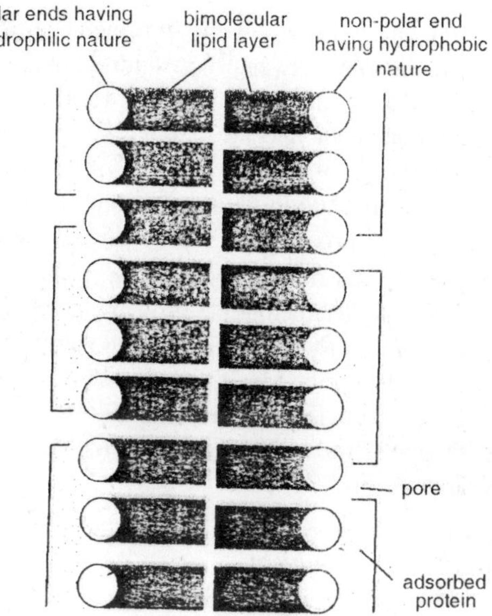

Fig. 2.2. Structure of plasma membrane.

The phospholipid bilayer of membrane also contains cholesterol. The membranes of mitochondria have just a few cholesterol molecules here and there, but some plasma membranes have as many cholesterols as they do phospholipids. Cholesterol affects membrane structure and function in several ways. It makes the bilayer stronger, more flexible but less fluid, and less permeable to water soluble substances such as ions or monosaccharides.

The flexible, somewhat fluid nature of the bilayer is very important for membrane function. As you breathe, move your eyes, and turn the pages of this book, cells in your body change shape. If their plasma membranes were stiff instead of flexible, cells would break open and die. Further, vesicles of membrane constantly bud off form the endoplasmic reticulum, merge with the Golgi complex, but off again, and fuse with the plasma membrane. Membranes can bud off vesicles and seal up again, or fuse with vesicles and smooth out again, because of the fluid nature of the bilayer.

Variety of Proteins in the Cell Membrane From a Protein Mosaic

A variety of proteins are embedded within or attached to the surface of a membrane's phospholipid bilayer. You will recall that some of the amino acids that make up proteins are hydrophilic whereas others are hydrophobic. Many membrane proteins are held in the lipid bilayer by the interactions between their hydrophobic amino acids and the hydrophobic tails of the phospholipids. Hydrophilic parts of the proteins may protrude either into the cytoplasm or into the extracellular fluid. Many of the proteins in plasma membranes also have carbohydrate groups attached to them, especially to the parts that stick outside the cell. These proteins are called *glycoproteins*.

Many proteins can move about within the relatively fluid phospholipid bilayer. Other membrane proteins, however, are anchored in place by a network of protein filaments connected to the cytoskeleton. In animal cells, which lack cell walls, the attachments between plasma membrane proteins and the underlying cytoskeleton produce the characteristic shapes of the cells, from the dimpled discs of red blood cells to the elaborate branching nerve cells.

Membrane proteins serve a variety of functions that allow the cell to interact with its environment

There are three major categories of membrane proteins, each of which serves a different function: transport proteins, receptor proteins, and recognition proteins.

Transport proteins regulate the movement of water-soluble molecules through the plasma membrane. Some, called *channel proteins*, form pores or channels that allow small water-soluble molecules to penetrate through the membrane. These proteins form a structure something like a sleeve with a lining: Hydrophobic amino acids (the outer material of the sleeve) anchor the protein in the phospholipid bilayer and hydrophilic amino acids form the inside of the channel (the lining of the sleeve). Every plasma membrane bears a large assortment of protein channels, each lined with specific amino acids that allow certain molecules, usually ions such as potassium (K^+), sodium (Na^+), and calcium (Ca^{2+}), to pass through. Other transport proteins, called *carrier proteins*, have binding sites, much like the active sites of enzymes, that can grab onto specific molecules on one side of the membrane. The transport protein then changes shape, in some cases through the use of cellular energy, and moves the molecule across the membrane.

Receptor proteins are molecular triggers that set off cellular responses when specific molecules in the extracellular fluid, such as hormones or nutrients, bind to them. Most cells bear dozens of different types of receptors on their plasma membranes. When activated by the appropriate molecule, some receptors set off elaborate sequences of cellular changes, such as increased metabolic rate, cell division, movement toward a source of nutrients, or secretion of hormones. Other receptors act like gates on channel proteins; activating the receptor opens the gates, allowing ions to flow through the channels. For example, these receptor-activated channels allow nerve cells in your brain to communicate with one another.

Recognition proteins and glycoproteins serve as identification tags and cell-surface attachment sites. The cells of your immune system, for example, recognize a *Salmonella* bacterium as a foreign invader and target it for destruction. These same immune cells ignore the trillions of cells of your own body because of different identification glycoproteins on their surfaces. During development, the growth of nerve fibers from your spinal cord down to the muscles in your feet is guided by attachments between recognition proteins on the nerve cell and the other cells it traverses on its way to the muscle.

As you can see from these brief descriptions, membrane proteins are largely responsible for the second, third, and fourth membrane functions in our list: moving substances across the membrane, communicating with other cells, and identifying the cell.

Transport Across Membranes

Nearly all living cells are bathed in liquid, which may be the extracellular fluid of a human body, the pond water in which the single-celled amoeba crawls, or the water-saturated cell walls of a young plant. The plasma membrane separates the fluid cytoplasm of the cell from its fluid environment. Let's begin our study of membrane transport, then, with a brief look at the characteristics of fluids, starting with a few definitions.

1. A *fluid* is a liquid or a gas; that is, any substance that can move or change shape in response to external forces without breaking apart.
2. The *concentration* of molecules in a fluid is the number of molecules in a given unit of volume.
3. A *gradient* is a physical difference between two regions of space, so that molecules tend to move from one region to the other. Cells frequently encounter gradients of concentration, pressure, and electrical charge.

Molecules in Fluids Move in Response to Gradients

The individual molecules in fluid move continually, bouncing off one another in random directions. Superimposed atop these random collisions, however, may be a very nonrandom *net movement* of particular types of molecules in the fluid, or even of the entire fluid, in response to gradients. Molecules move from regions of high concentration to low concentration (for example, sugar dissolving in coffee) or from high pressure to low pressure (such as air flowing out of a bicycle pump when you push down the handle). Charged molecules (ions) move in response to electrical gradients, toward unlike charged or away from like charges. The gradients, toward unlike charges or away from like charges. The gradient of concentration, pressure, or electrical charge provides the potential energy that can drive the net movement of the molecules. By analogy with gravity, we will refer to such movements as going "down" the gradient.

Movement across Membrane Occurs by Both Passive and Energy-Requiring Transport

Because the cytoplasm of a cell is very different from the extracellular fluid, gradients of concentration, electrical charge, and occasionally pressure span the plasma membrane. In its role as gatekeeper of the cell, the plasma membrane provides for two types of movement.

During *passive transport*, substances move down gradients of concentration, electrical charge or pressure. This movement by itself requires no expenditure of energy. The gradients provide the potential energy that drives the movement and controls the direction of movement, into or out of the cell. The plasma membrane acts as a filter; the lipids and protein pores regulate which molecules can cross and may influence the rate and timing of movement, but they do not influence the direction of movement.

In energy-requiring transport, substances move against gradients of concentration, electrical charge, or pressure. In this case, transport proteins in the plasma membrane do control the direction of movement, using chemical energy from cellular metabolism to drive movement against the gradients. A helpful analogy for understanding the differencc between these types of transport is to consider what happens when you ride a bike. If you don't pedal, you can go only downhill, as in passive transport. However, if you put enough energy into pedaling, you can go uphill as well.

Passive Transport: Movement down Concentration Gradients

Although gradients of pressure and electrical potential are important in regulating the movement of certain substances in living organisms, most materials that move passively across plasma membranes do so in response to *concentration gradients*.

Molecules Diffuse from Areas of High Concentration to Area of Low Concentration

Diffusion is the net movement of molecules in a fluid from regions of high concentration to regions of low concentration, driven by the concentration gradient. Diffusion can occur from one part of a fluid to another or across a membrane separating two fluid compartments. We will first examine the simplest case, the diffusion of molecules within a fluid.

To see how concentration gradients cause molecules to diffuse from one place to another, let's consider what happens when a drop of dye is placed in a glass of water. With time, the drop will seem to become larger and paler, until eventually, even without stirring, the whole glass of water will be uniformly faintly colored. Why?

A drop of dye consists of many individual dye molecules moving in all directions. Those molecules that move but stay within the drop don't change its composition, which was all dye to begin with. But

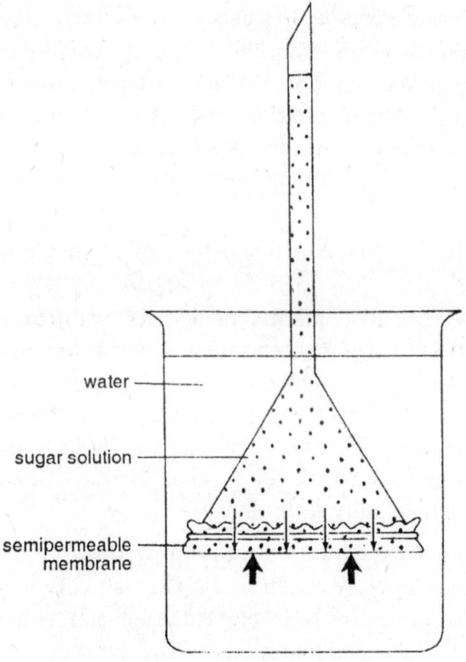

Fig. 2.3. Simple membrane osmometer.

some molecules along the border of the drop, simply due to random motion, go out into the water. Thus, the net movement of dye molecules is from the region of high dye concentration (the water). The same thing happens with water molecules. Random motion causes some to intersperse with the dye molecules, and the net movement of water is from the high water concentration outside the drop into the low water concentration inside the drop.

To the observer, dye molecules that move beyond the original border of the drop make the drop appear larger; water molecules that invade the drop dilute the dye, making the drop paler. At first, when the drop is pure dye and the water is pure water, there is a very steep concentration gradient, and the dye diffuses rapidly. As the concentration gradient, and the dye diffuses rapidly. As the concentration differences lessen, the dye diffuses more and more slowly. However, as long as the concentration of dye within the expanding drop is greater than the concentration of dye in the rest of the glass, the net movement of dye will be from drop to water, until the dye becomes uniformly dispersed in the water. Then, with no concentration gradient of either dye or water, diffusion stops. Individual molecules still move about, but there are not changes in concentration of either water or dye.

As you can appreciate from this imaginary experiment, diffusion cannot move molecules rapidly over long distances. Although the drop of dye immediately begins to diffuse into the water, uniform dispersion may take many minutes. As described in the previous chapter, the slow rate of diffusion over long distances is one of the reason cells are small.

Molecules Diffuse across Membranes down Their Concentration Gradients

Many molecules cross plasma membranes by diffusion, driven b differences between their concentration in the cytoplasm and in th external environment. Molecules cross the plasma membrane at differe locations and at different rates, depending on the properties of tl molecule in question. Therefore, plasma membranes are said to *differentially permeable*: They allow some molecules to pass throu or *permeate*, more rapidly than others.

Some molecules move across membranes by simple diffusion

Water, dissolved gases such as oxygen and carbon dioxide, d lipid-soluble molecules such as ethyl alcohol and vitamin A eay diffuse across the phospholipid bilayer. This process is called *sile diffusion*. Generally, the rate of simple diffusion is a function ofne

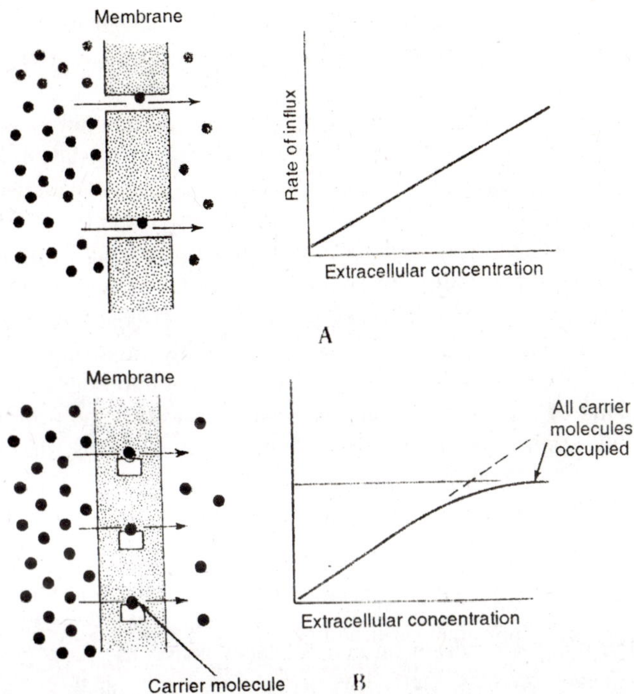

Fig. 2.4. (A) Diffusion through membrane channels. (B) Carrier-mediated transport.

concentration gradient across the membrane, the size of the molecule, and how easily it dissolves in lipids (its lipid solubility): Large concentration gradients, small molecule size, and high lipid solubility all increase the rate of simple diffusion.

Other molecules cross the membrane by facilitated diffusion with the help of membrane proteins

Most water-soluble molecules, such as ions (K^+, Na^+, Ca^{2+}), amino acids, and monosaccharides, cannot move through the phospholipids bilayer. These molecules can diffuse across only with the aid of two types of transport proteins, channel proteins and carrier proteins, which we discussed earlier in this chapter. This process is called *facilitated diffusion*.

Channels proteins from permanent pores, or channels, in the liquid bilayer through which certain ions can cross the membrane. Usually, each channel protein has a specific interior diameter and distribution of electrical charge that allow only particular ions to pass through. Nerve cells, for example, have separate channels for sodium ions, potassium ions, and calcium ions. Many of these channels have gates on one end, which open and close in response to electrical or chemical signals.

Carrier proteins bear groups of amino acids that bind specific molecules from the cytoplasm or extracellular fluid. Binding triggers a change in the shape of the carrier that allows, the molecule to pass through the protein and thus across the plasma membrane. Carrier proteins that do not use cellular energy move molecules only down their concentration gradients.

Plasma membranes contain relatively few channels or carriers through which any given molecule can pass. The rate of facilitated diffusion is therefore a function of both the concentration gradient and the amount of transport proteins. Molecules that cross the membrane by facilitated diffusion usually do so more slowly than those that cross by simple diffusion through the lipid bilayer.

Osmosis is the Diffusion of Water across Membranes

Water, like any other molecules, moves by diffusion from regions of high water concentration to regions of low water concentration. However, the diffusion of water across differentially permeable membranes has such dramatic consequences that it has been given a special name: *osmosis*. Let's investigate osmosis in another thought experiment. A very simple kind of differentially permeable membrane consists of an impervious sheet perforated with tiny pores. The pores allow water molecules to pass through, but not larger molecules such as sugar. Suppose we make a bag out of such a membrane, fill it with a sugar solution, tie off the top, and place the bag in a glass of pure water. The bag will swell up, and, if it is weak enough, it will burst. Why?

If you could see individual molecules, you wild notice that there are two categories of water molecules in the sugar solution inside the bag: "free" water molecules, well separated from the sugars; and "bound" water molecules, held to the sugars by hydrogen bonds (This is why sugars dissolve in water). In the pure water outside the bag, of course, there are only free water molecules. Now, free water molecules can diffuse through the pores in the membrane, but the bound water molecules cannot, because they are attached, at least temporarily, to the bulky sugars.

Therefore, the concentration of free water molecules is lower inside the bag than in the pure water outside. This water concentration gradient favors the movement of free water molecules from the pure water outside the bag to the sugar solution inside the bag. The bag swells up as more water molecules enter the bag than leave it. The sugar can't escape at all, so the free water concentration inside the bag is always lower than in the pure water outside. Water continues to enter the bag until it bursts.

Now let's redo our experiment but with one change. Suppose we tie a piece of the same differentially permeable membrane across the end of a glass tube, suspend the tube in pure water, and put the sugar solution in the tube. Water will diffuse across the membrane from the pure water into the solution in the tube. Because the upper and of the tube is open, as the water enters it will raise the level of the solution in the tube. Eventually, the solution in the tube will stop rising. Why?

As the solution in the tube rises above the level of water in the glass, its weight applies a pressure to the membrane across the bottom of the tube. Therefore, two opposing gradients are set up: a concentration gradients moving water into the tube and a pressure gradient pushing water out of the tube. When the two gradients are equal, no further net movement of water will occur across the membrane. The physical pressure that exactly balances the osmosis of water due to the concentration difference between a solution and pure water is defined as the *osmotic pressure* of the solution. A solution with a high osmotic pressure has a low concentration of free water; a solution with a low osmotic pressure has a high concentration of free water.

As we have emphasized, osmosis is driven by differences in *water concentration* between solutions. Therefore, osmosis does not depend on the type of *dissolved molecules* in a solution, but only on their *concentration*. We can produce the same osmotic pressure by adding sucrose, glucose, amino acids, sodium ions, or a mixture of all of these to a solution, as long as the concentration of all types of particles that cannot permeate a membrane stays the same.

Osmosis across the Cell Membrane Plays an Important Role in the Lives of Cells

Most plasma membranes are highly permeable to water. Because all cells contain dissolved salts, proteins, sugars, and so on, the flow of water across the plasma membrane depends on the concentration of water in the liquid that bathes the cells. The extracellular fluids of animals are usually *isotonic* ("having the same strength") to the inside of the body cells; that is, the concentration of water inside is the same as that outside the cells so there is not net tendency for water to either enter or leaves the cells. Note that the types of dissolved particles are seldom the same inside and outside the cells, but the total concentration of all dissolved particles is equal, with the result that the water concentration inside is equal to that outside the cells.

If red blood cells, for example, are taken out of the body and immersed in salt solutions of varying concentrations, the effects of the differential permeability of the plasma membrane to water and dissolved particles become dramatically apparent. If the solution has a higher salt concentration than the cytoplasm (that is, if the solution has a lower water concentration), water will leave the cells by osmosis. The cells will shrivel up until the concentration of water inside and outside become equal. Solutions that cause water inside and outside become equal. Solutions that cause water to leave by osmosis are called *hypertonic* ("having greater strength").

In the opposite situation, if the solution has little or no salt, water will enter the cells, causing them to swell. If the solution has little enough salt, the cells will burst. Solutions that cause water to enter by osmosis are called *hypotonic* ("having lesser strength").

Osmosis across plasma membranes is crucial to the functioning of many biological systems, including water uptake by plant roots, absorption of dietary water from your intestines, and reabsorption of water and minerals in your kidneys. Osmosis will occur across any membrane, including the membranes surrounding organelles within a cell, if the water concentration differs on the two sides of the membrane. Osmosis across internal membranes, such as those that surround vacuoles, can also be important in the lives of certain cells.

Energy-Requiring Transport across Membranes

All cells need to move some materials "uphill" across their plasma membranes, against diffusion gradients. For example, every cell requires some nutrients that are less concentrated in the environment than in the cell cytoplasm. Therefore, diffusion would cause the cell to lose, not gain, these nutrients. Other substances, such as sodium and calcium ions in your brain cells, must be maintained at much lower concentrations inside the cells, must be maintained at much lower concentrations inside the cells than in the extracellular fluid. When these ions diffuse into the cells, they must be pumped out again against their concentration gradients. Finally, many cells acquire or expel some substances, such as whole bacteria or large proteins, that are too large to diffuse across a membrane regardless of concentration gradients. Cells have evolved several processes that use cellular energy to move materials into or out of the cell.

Active Transport Uses Energy to Move Molecules across Membranes

In *active transport*, membrane proteins use cellular energy to move individual molecules across the plasma membrane, usually against their concentration gradient. Active transport proteins span the membrane and have two active sites. One active site (which may be either on the face of the plasma membrane in contact with the cytoplasm or on the face in contact with the extracellular fluid, depending on the active transport protein) recognizes a particular molecule, say a calcium ion, and binds it. The second site (always on the inside of the membrane) binds an energy-carrier molecule, usually adenosine triphosphate (ATP). The ATP donates energy to the protein, causing it to change shape

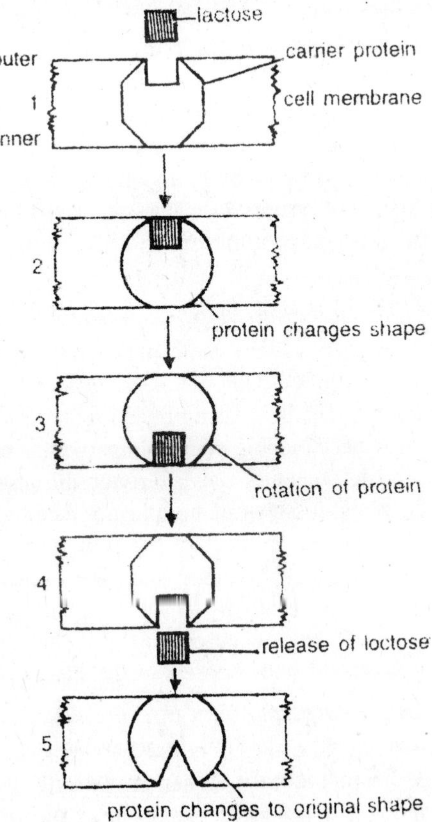

Fig. 2.5. Diagram showing active transport based on revolving door model.

and move the calcium ion across the membrane. Active transport proteins are often called *pumps*, in analogy to water pumps, because they use energy to move molecules "uphill" against a concentration gradient. We will see that plasma membrane pumps are vital in mineral uptake by plants, mineral absorption in your intestines, and maintaining concentration gradients essential to nerve cell functioning.

Cells Engulf Particles or Fluids by Endocytosis

Cells can acquire fluids or particles, especially large proteins or entire microorganisms such as bacteria, by a process called *endocytosis* (Greek for "into the cell"). During endocytosis, the plasma membrane engulfs the particle of fluid droplet and pinches off a membranous sac called a *vesicle*, with the particle or fluid inside, into the cytoplasm. Three types of endocytosis can be distinguished on the basis of the size of the particle acquired and the method of acquisition.

Pinocytosis moves liquid into the cell

In *pinocytosis* ("cell drinking), or *fluid-phase endocytosis*, a very small patch of plasma membrane dimples inward as it surrounds extracellular fluid and buds off into the cytoplasm as a tiny vesicle. Pinocytosis moves a droplet of extracellular fluid, contained within the dimpling patch of membrane, into the cell. Therefore, the cell acquires materials in proportion to their concentration in the extracellular fluid.

Receptor-mediated endocytosis moves specific molecules into the cell

Cells can take up certain molecules (cholesterol, for example) more efficiently by a process known as *receptor-mediated endocytosis*. Most plasma membranes bear many receptor proteins on their outside surface, each bearing a binding site for a particular nutrient molecule. In most cases, these receptors move through the phospholipid bilayer and accumulate in depressions of the plasma membrane called coated pits. If the right molecule contacts a receptor protein in one of these coated pits, it attaches to the binding site. The coated pit deepens into a U-shaped pocket that eventually pinches off into the cytoplasm as a coated vesicle. Both the receptor-nutrient complex and a bit of extracellular fluid move into the cell in the coated vesicle.

Phagocytosis moves large particles into the cell

Phagocytosis ('cell eating") is used to pick up large particles, including whole microorganisms. When an *Amoeba*, for example, sense a tasty *Paramecium*, it produces extension of its surface membrane, called *pseudopodia* (Latin for "false foot'; singular, *pseudopod*). The pseudopodia surround the luckless *Paramecium*, their ends fuse, and the prey is carried into the interior of the *Amoeba* for digestion. The resulting vesicle, called a food vacuoles, fuses with lysosomes whose enzymes digest the prey. While blood cells also use phagocytosis and intracellular digestion to engulf and destroy bacteria that have invaded your body.

Exocytosis moves material out of the cell

The reverse of endocytosis, called *exocytosis* (Greek for "out of the cell"), is often used by cells to dispose of unwanted materials, such as the waste products of digestion, or to secrete materials, such as hormones, into the extracellular fluid. During exocytosis, a vesicle created by the Golgi complex moves to the cell surface, where the membrane of the vesicle fuses with the plasma membrane. The vesicle opens to the extracellular fluid and its contents diffuse out.

Transport across Intracellular Membranes

As we mentioned earlier, all the membranes of a cell, including those that surround such diverse organelles as chloroplasts, mitochondria, the nucleus, and vacuoles, are similar in structure and function to the plasma membrane. The transport processes that we have described in this chapter, form simple diffusion to active transport, also occur across the internal membranes of a cell. For example, chloroplasts and mitochondria have specific protein channels that are permeable to hydrogen ions and that are crucial to ATP production in those organelles.

Calcium ions are actively transported across certain internal membrane in muscle cells. And if you look back at the section on membrane flow within cells in chapter 5, you will see vesicle budding from the fusing with the membranes of the endoplasmic reticulum and Golgi complex, processes virtually identical to endocytosis and exocytosis. Transport across internal membranes is just as vital to the life of a cell as is transport across the plasma membrane, and similar mechanisms are used, as illustrated by transport in and out of cell vacuoles.

Vacuoles Help Regulate Cell volume

Most cells contain one or more vacuoles, which are fluid-filled sacs surrounded by a single membrane. Some, such as the food vacuoles that form during phagocytosis, are temporary features of cells. Many cells, however, contain permanent vacuoles that have important roles in maintaining the integrity of the cell, especially by regulating the cell's water content.

Contractile vacuoles are found in freshwater microorganisms

Freshwater protists such as Paramecium often contain complex *contractile vacuoles* composed of collecting ducts, a central reservoir, and a tube leading to a pore in the plasma membrane. Because fresh water is hypotonic to the cytosol of these organisms, water constantly enters the cell by osmosis. The increasing volume of incoming water might soon burst the fragile creature if it did not have a mechanism to excrete the water. Water is pumped into the collecting ducts and drains into the central reservoir (a process that require cellular energy). When the reservoir is full, it contracts, squiring the water up the exit tube and out through the pore in the plasma membrane.

Central vacuoles are found in plant cells

Three-quarters or more of the volume of many plant cells is occupied by a large *central vacuole*. The central vacuole has several functions. It provides a dump site for hazardous wastes, which plant cells often cannot excrete. Some plant cells store extremely poisonous substances, such as sulfuric acid, in their vacuoles; these substances deter animals from munching on the otherwise tasty leaves. Vacuoles may also store sugars and amino acids not immediately needed by the cell. Blue or purple pigments stored in central vacuoles are responsible for the colors of many flowers.

These dissolved substances make the vacuole contents hypertonic to the cell cytoplasm, which in turn is usually hypertonic to the extracellular fluid bathing the cells. Water therefore enters the vacuole by osmosis, which tends to make it swell. The pressure of the water within the vacuole, called *turgor pressure*, pushes the cytosol up against the cell wall with considerable force. Primary cell walls are usually somewhat flexible, so the overall shape and rigidity of the cell depend on turgor pressure within the cell. Turgor pressure thus provides support for the nonwoody parts of plants. If you forget to water your houseplants, the central vacuoles and cytosol lose water and the cell shrinks away from its cell walls. Just as a balloon goes limp when its air leaks out, so too the plant droops as its cells lose turgor pressure.

Cell Connections and Communication

In multicellular organisms, plasma membranes also function in holding together clusters of cells and in providing avenues through which cells communicate with their neighbour. Depending on the organism and the cell type, one of four types of connection may occur between cells: desmosomes, tight junctions, gap junctions, and plasmodesmata.

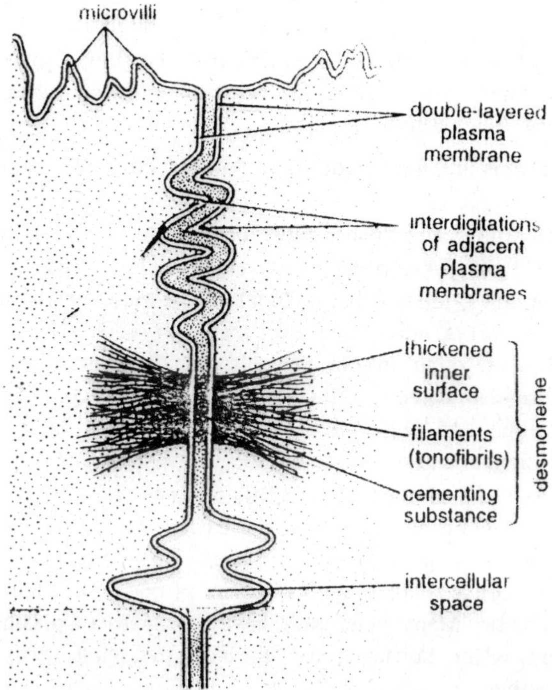

Fig. 2.6. Portions of two adjacent cells, showing intercellular space, a desmosome and a region of interdigitation of plasma membrane.

Desmosomes Attach Cells Together

Animals as you know, tend to be flexible, mobile organism. Many of an animal's tissues are stretched, compressed, and bent as the animal moves about. If the skin, intestines, stomach, urinary bladder, and other organs are not to tear apart under the stresses of movement, their cells must adhere firmly to one another. Such animals tissues have junctions called *desmosomes* that hold adjacent cells together. In a desmosome, the membranes of adjacent cells are glued together by proteins and carbohydrates. Intermediate filaments attached to the inside of the desmosomes extend into the interior of each cell, further strengthening the attachment.

Tight Junctions Leakproof the Cell

The animal body contains many tubes or sacs that must hold their contents without leaking: A leaky urinary bladder would spell disaster for the rest of the body. The spaces between the cells lining such sacs are sealed with strands of protein to form *tight junctions*. The membranes of adjacent cells nearly fuse along a series of ridges, effectively forming waterproof gaskets between cells. Continuous tight junctions sealing each cell to its neighbors prevent molecules from escaping between cells.

Gap Junctions and Plasmodesmata Allow Communication between Cells

Multicellular organisms must coordinate the actions of their component cells. In animals, many cells, including their component cells. In animals, many cells, including heart muscle cells, most gland cells, some brain cells, and every cell of very young embryos, communicate through protein channels directly connecting the insides of adjacent cells. These cell-to-cell channels are clustered in specialized regions called *gap junctions*. Hormones, nutrients, ions, and even electrical signals can pass through the channels at gap junctions.

Virtually all of the living cells of plants are connected to one another by much larger channels called *plasmodesmata*. Each plasmodesma is a tube, lined with plasma membrane, that penetrates through the cell wall from the cytoplasm of one cell to the cytoplasm of its neighbor. Many plant cells have thousands of plasmodesmata. As a result, water, nutrients, and hormones pass quite freely from one cell to another.

Natural Selection Has Produced Cell Membranes with Diverse Compositions

Although the membranes of all cells have a similar structure, membrane function varies tremendously from organism to organism and from cell to within a single organism. This diversity arises largely from the different proteins and phospholipids in the membrane, which have evolved under differing selection pressures.

Our discussion of membranes emphasized the unique functions of the membrane proteins. Consequently, you may think that the phospholipids are just a waterproof place for the proteins to reside. This isn't quite true, as we can see by examining the plasma membrane phospholipids in legs of caribou, animals that live in very cold regions of North America. During the long arctic winters of these regions, temperatures plummet far below freezing, and for caribou to keep their legs and feet really warm would waste precious energy. These conditions have favored the evolution of specialized arrangements of arteries and veins in caribou legs that allow the temperature of their lower legs to drop almost to freezing (0°C). The upper legs and main trunk of the body, in contrast, remain at about 37°C. Further, the phospholipids in the membranes of cells in the upper legs of caribou are very different from those near the hooves.

Remember, the membrane of a cell needs to be somewhat fluid to allow the proteins to move to sites where they are needed. The fluidity of membrane is a functions of the fatty acid tails of its phospholipids: Unsaturated fatty acids remain more fluid at lower temperatures than saturated fatty acids do. Caribou have a range of phospholipids in the plasma membranes of the cells in their legs. The membranes of cells near the chilly hoof have lots of unsaturated fatty acids, whereas the membranes of cells near the warmer trunk have more saturated ones. This arrangement give the plasma membranes throughout the leg the proper fluidity despite great differences in temperature.

As important as the phospholipids are, the membrane proteins probably play the major roles in determining cell function and in governing the interactions between a cell and its neighbors. The extreme complexity of the protein molecule makes it more susceptible than other types of molecules to mutations that alter its amino acid composition, its shape, and hence its function. Over billions of years, an incredible diversity of proteins has evolved. Every nerve cell in your body, for instance, has membrane proteins essential for producing electrical signals and conducting them along the nerves to various parts of the body. Other membrane proteins receive chemical messages from neighboring nerve cells or from hormones and other chemicals in the blood. Each cell in the brain has a specific set of membrane proteins, allowing it to respond to some stimuli while ignoring others. In fact, your ability to read this page depends on the proteins residing in the membranes of your brain cells.

3

Golgi Complex

In 1898 by means of a silver staining method, *Golgi* discovered a reticular structure in the cytoplasm. The name "Golgi apparatus," generally given to this structure, is confusing because it suggests a definite relationship with the physiologic processes of the cell. Today it seems more appropriate to use the name "*Golgi substance*" or "*Golgi complex*," to refer to this material that has special staining properties. Because its refractive index is similar to that of the matrix, the Golgi complex is difficult to observe in living cells. The use of the electron microscope has provided a distinct image of this component, and its submicroscopic structure has been revealed.

For years the Golgi complex was thought to be an artifact of various fixation and staining precedures. In other words, many scientists believed that the structure observed during numerous microscopy procedures and termed the Golgi did not actually exist in the living cell. *Guilliermond* (1923), *Parat* (1927) *Walker* and *Allen* (1921), doubts arose with regard to the existence of Golgi complex. The different lines of objections against the existence of Golgi complex are as follows:

1. Nothing corresponding to the Golgi elements exists in the living cells, what appears Golgi bodies in the processed cells are myeline figures, produced by the action of cytological reagents or lipids *Walker* and *Allen* (1921); *Palade* and *Claude* (1949), however, believe in the existence of Golgi bodies in the germ cells but no such things in the somatic cells.
2. The Golgi apparatus of the fixed material is produced by the deformation and metallic impregnation of several unrealed cytoplasmic bodies.

3. The dictyosome or acroblasts of male germ cells homologised with the Golgi apparatus of nerve cells. *Gatenby, Bowen* and *Hirschler, Parat* (1928) believed that they are nothing but the modified mitochondria.
4. The Golgi apparatus is a stacks of the smooth endoplasmic reticulum.

Answer to the Objections

1. In a number of cells these bodies have been studied in *vivo* or in *vitro* with phase contrast microscopy.
2. The Golgi bodies have been separated by centrifugal fractionation of homogenized cell.
3. The objection of artifact is discarded on the basis that artifact cannot show the pattern of the same type as shown by these bodies.
4. Their ultra structure has been recently studied with the electron microscope by *Sjostrand* and *Hanzon* (1954) and *Dalton* and *Felix* (1954).
5. Dictyosomes of male germ cells which are impregnated with silver and osmium, have same type of chemical composition as the Golgi apparatus of somatic cells.
6. According th *Srivastava* (1953, 1954) dictysomes in male germ cells, are favourable materials for the study of Golgi bodies even under the light microscope.
7. The idea that the dictyosomes are modifed mitochondria has been discarded on the basis of biochemical tests. (a) The dictyosomes of male germ cells contain appreciable quantity of polysaccharides. It is also present in the somatic Golgi bodies but not found in mitochondria. Similarly alkaline phosphatase is associated with Golgi bodies, but not found in the mitochondria. (b) *Tiwari* and *Broune* (1962, 1963) have found that the distribution of adenosine triphosphatase is localized in mitochondria but not in Golgi bodies. In the same way succinic dehydrogenase and cytochrome oxidase are noted from the mitochondria, but not in the Golgi bodies.
8. The statement that the Golgi bodies of neuron are nothing but endoplasmic reticulum, is discarded on the basis of electron microscopic structure and the width of the Golgi membrane is different from that of the endoplasmic reticulum.

In concluding the controversy of the Golgi apparatus, it should be pointed out that *Baker* (1963) and his students were the last to accept the reality of the Golgi apparatus when *Baker* wrote, "The author accepts, after long hesitation, the view that the Golgi apparatus in the neurons of vertebrates with the organelle of the same name, in other cells."

Terminology of Golgi Complex

Holmgren referred to the Golgi complex as *trophospongium* (*Cajal* referred it as the *Golgi-Holmgren canals*). *Baker* used the term *lipochondria* because of the presumed lipid content. The term *Dalton Complex* was given after the name of its observer *Dalton* in 1952. *Sjostrand* proposed the term cytomembranes for the Golgi system.

Sosa has suggested the following nomenclature for the Golgi Complex.

1. *Golgiokinesis.* Division of the Golgi apparatus during nuclear division.
2. *Golgiosomes.* Corpuscles produced by the Golgiogenesis are called as Golgiosomes which are described as Gogli material in invetebrates.
3. *Golgiolysis.* Process of dissolution of the Golgi apparatus.
4. *Golgiorrhexis.* Fragmentation of the Golgi apparatus.
5. *Golgiogenesis.* Formation and differentiation of the Golgi body during embryonic development.
6. *Golgi-cytoarchitecture.* Study of structure of cell in relation to Golgi apparatus.

Occurrence

The Golgi complex occurs in all cells except the prokaryotic cells (viz., Mycoplasma, bacteria and blue green algae) and eukaryotic cells of certain fungi, sperm cells of bryophytes and pteridophytes, cells of mature sieve tubes of plants and mature sperm and red blood cells of animals. Their number per plant cell can vary from several hundred as in tissues of corn root and algal rhizoids (i.e., more than 25,000 in algal rhizoids, *Sievers,* 1965), to single organelle in some algae. In animal cells, there usually occurs a single Golgi complex, however, developing oocytes of chordates may contain many Golgi complex.

Morphology of Golgi Complex

The morphology of the Golgi complex varies from cell to cell depending uupon the type of cell in which they are found. Two forms of Golgi complex have been observed.

1. *Localized form.* In polarised cells of vertebrates (which have base and apex). Golgi complex occurs singly and occupies a fixed

position. It lies between the nucleus and secretory pole. This can be best seen in thyroid cells, in exocrine cells of pancreas and the mucous cells of intestine.

2. *Diffused form.* In some specialized cells of vertebrates (nerve cells and liver cells), in most plant cells and in the cells of invertebrates several units of Golgi complex are found scattered along with the elements of endoplasmic reticulum. Each unit is called a *dictyosome*. In liver cells these occur as many as 50 dictyosomes per cell and in certain plant cells their number may reach upto hundreds.

Shape

The shape of the Golgi complex is quite variable in different somatic cell types of animals. Even in the same cell there are variations in different functional stages. The shape is, however, constant with each cell type. It varies in form from a compact mass to a disperse filamentous network.

Number

The number of Golgi stacks per cell varies enormouly, depending on the cell type—from as few as one to the hundreds. There is a

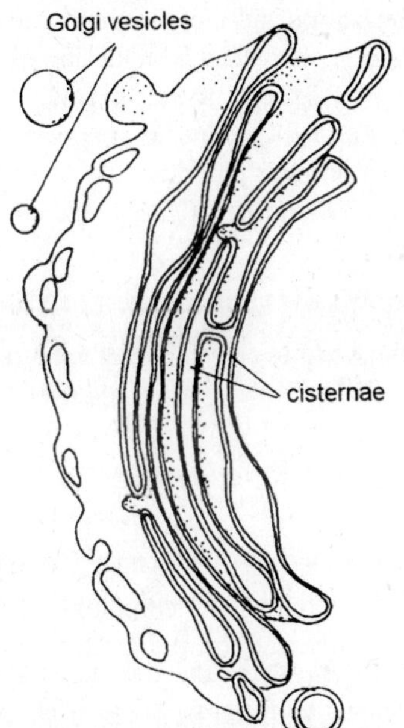

Fig. 3.1. Diagram showing the structural organization of a Golgi body.

single large in some cells while in the case of *Paramoeba* there are two. In *Stereomyxa* (a species of *Amoeba*) there are many Golgi complexes. Nerve cells, liver cells and most plant cells also have multiple Golgi complexes, there being about 50 in liver cells. The Golgi complex can even account for a large fraction of the cell volume in some specialized cells. One example is the goblet cell of the intestinal epithelium, which secretes mucus into the gut; the glycoproteins in mucus are glycosylated principally in the Golgi complex.

Size

The size is likewise very variable. It is large in the nerve and gland cells and small in the muscle cells. In general the Golgi complex is well-developed while the cell is in active state. When the cell grows old, the complex progressively diminishes in size and disappears.

Position

The position of the Golgi complex is relatively fixed for each cell type. In the cells of ectodermal origin, the Golgi complex is polarized from the time of the embryonic state between the nucleus and the periphery (*Cajal,* 1914). In the secretory exocrine cells that have in general a typical polarization the Golgi complex is found between the nucleus and the secretory pole. In the endocrine glands the polarity of this organoid is variable, except in the thyroid, where it is oriented towards the centre of the follicle. In the yonger cells and often in the older ones it lies most commonly at one side of nucleus but in certain cases may completely surround it. In ganglionic cells of the mouse the position in perinuclear.

Detailed Structure of Golgi Complex

Dalton and *Felix* (1954) described the Golgi complex in the epididymis of the rat after taking the first electron micrographs. The following description of the Golgi complex is a composite one based on the work of several authors.

Cisternae

The cisternae or saccules are similar to the smooth surface ER, and appear in section as stacks of closely spaced membrane-delimited sacs. The number of saccules varies from about 4 to 8 in most animal and plant cell types. In *Euglena*, the number may go upto 20. The membrane of the saccule is roughly 60 to 70Å in thickness which encloses a cavity about 150Å wide whose edges are often dilated. According to most authors, there are two well-defined faces of the

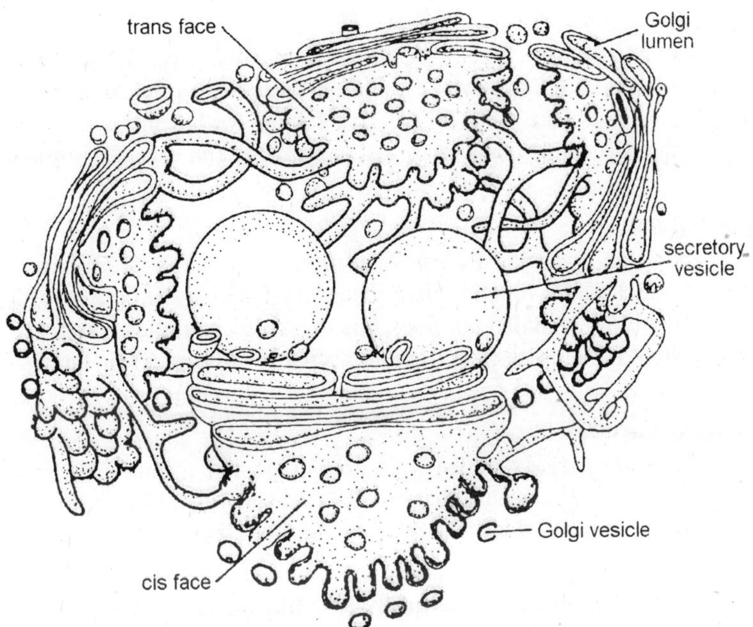

Fig. 3.2. Three-dimensional drawing of a Golgi apparatus; it drawn from electron micrographs of a secretory animal cell.

cisternae, i.e., convex and the concave; the later is generally referred to as the mature or forming or distal face and the convex side is assumed to be the immaure or excit or proximal face, the cisternae lie in parallel array are separated from each other by a space of about 200 to 300Å. What holds them together is not yet known but in few cells a thin layer of electron opaque, sometimes dense material is seen between the saccules which at certain regions are more prominent to which *Amos* and *Grimstone* (1968) applied the term nodes. *Mollenhauer et al.,* (1973) explored in some detail intercisternal elements and plaques in certain plant Golgi complex.

Tubules

From the peripheral area of cisternae arise a complex, anastomosing flat network of tubules of 300 to 500Å diameter. *Clowes* and *Juniper* (1969) have compared this tubular network to disc of lace.

Vesicles

The vesicles are small droplet-like sacs which remain attached to tubules at the periphery of the cisternae. They are of following two types:

Smooth vesicles

The smooth vesicles are of 20 to 80 mμ diameter. They conain secretory material (so often are called secretory vesicles) and are budded off from the ends of cisternal tubules within the net. Often more than one tubule connect to, and presumably fill, a single forming vesicle.

Coated vesicles

The coated vesicles are spherical protuberances, about 50 mm in diameter and with a rough surface. They are found at the periphery of the organelle, usually at the ends of single tubules, and are morphologically quite distinct from the secretory vesicles. Their function is unknown.

Golgian Vacuoles

These are large rounded sacs present on the maturing face of Golgi. These are formed either by the expanded cisternae or by the fusion of secretory vesicles. The vacuoles are filled with some amorphous or granular substance.

The Golgi Complex is Structurally and Biochemically Polarized

The Golgi complex has two distinct faces: a *cis* or *forming face* and a *trans,* or *maturing face*. The cis face is closely associated with a smooth transitional portion of the rough ER. In secretory cells, the trans face is the face closest to the plasma membrane; here, the large secretory vesicles are found exclusively in association with the trans face of a Golgi stack, and the membrane of a forming secretory vesicle is often continuous with that of the trans face of the last ("trans-most") cisterna. In contrast, the small Golgi vesicles are localized more evenly along the stack, proteins are commonly thought to enter a Golgi stack from the ER on the cis side and to exit for multiple destinations on the trans side; however, neither their exact path through the Golgi complex nor how they travel from cisterna along each stack are known.

The two faces of the Golgi complex are biochemically distinct. For example, a variation in the thickness of the Golgi membranes can be detected across the stack in certain cases, with those at the cis side being thinner (ER-like) and those at the trans side being thicker (plasma membrane like). More striking are the results obtained when certain histochemical tests are used in conjunction with electron microscopy to localize particular proteins within the Golgi complex. Some of these tests reveal membrane-bound enzyme activities that show a distinct polarity in their localization within the Golgi stack. A particularly intriguing biochemical finding was the discovery that lysosomal enzymes, such as acid phosphatase, are concentrated with the trans-most cisterna of the Golgi stack and within some of the coated vesicles nearby. This suggests that specific vesicles leaving for lysosomes are assembled in this region.

Secretory proteins are found by histochemical methods in all of the stacked cisterna, even though the large secretory vesicles in which these products are concentrated and associated only with the trans-most Golgi cisterna.

Chemical Composition

Regarding the chemical composition of Golgi complex, it has been demonstrated that the following substances are present:

Phospholipids

Phospholipids composition of Golgi membranes is intermediate between those of endoplasmic membranes and plasma membranes.

Proteins and Enzymes

Golgi complex from different plant and animal cells show great variations in the protein and enzyme contents. Some of the enzymes are ADPase, ATPase, NADPH cytochrome-C-reductase, glycosyl transferases, galactosyl transferase, thiamine pyrophosphate etc.

Carbohydrates

Both plant and animal cells have some common carbohydrate components, like glucosarine, galactose, glucose, mannose, and fructose. Plant Golgi lack sialic acid, but it occurs in high concentration in rat liver. Some carbohydrates like xylase and arbinose are present in plant cells only.

Vitamin C

The function of vitamin C stored in the Golgi complex has been shown by *Tomitte*. According to him Golgi complex stores vitamin C and liberates it slowly into the cytoplasm in sufficient amount of prevent-oxidation of the cell products.

FUNCTION OF GOLGI COMPLEX

Formation of Acrosome during Spermiogenesis

During the maturation of sperm the Golgi complex plays a role in the formation of acrosome.

In early stages, the Golgi appears as a spherical body, comprising cisternae arranged in parallel stacks and numerous small vesicles.

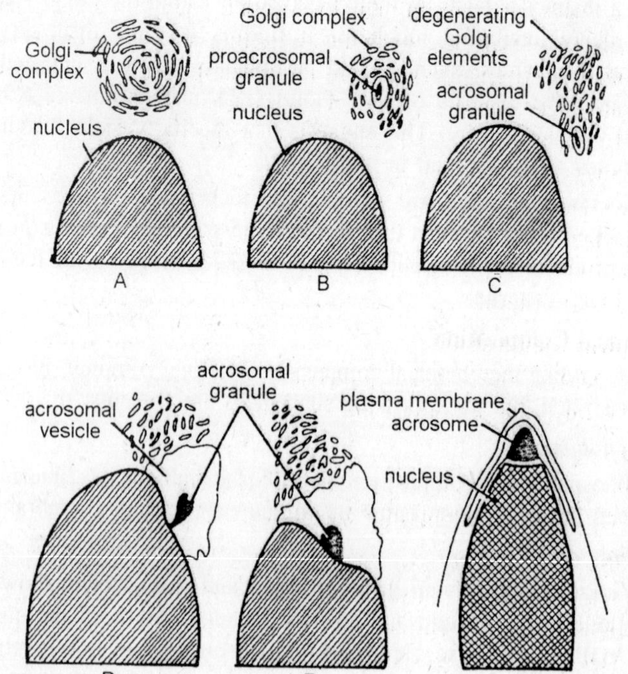

Fig. 3.3. Diagram illustrating gradual modification in Golgi complex to form acrosome during spermiogenesis.

The latter always pinched off from the cisternae. As development proceeds, the Golgi complex becomes irregular in shape and large vacuoles are formed by dilations of cisternal sac. In the centre of these large vacuole or vacuoles is present a dense granule, the *proacrosomal granule*. This granule which is derived from Golgi complex continues to grow within the vacuole by a process known as *accretion*. This vacuole and granule approaches the anterior pole of the nuclear membrane, constituting *acrosomal granule*. With the elongation of the spermatid, the *acrosomal vesicle* spreads over the nuclear surface and finally collapses with the nuclear membrane, forming the *cap material*. The acrosomal granule becomes the *acrosome* which lies at the apex of the nucleus and apparently comprises certain enzymes involved in the process of fertilization.

Synthesis and Secretion of Polysaccharides

Studies on goblet cells by autoradiography and electron microscopy have established the inter-relationship between protein synthesis, carbohydrate addition and sulphation. The goblet cells of the colon

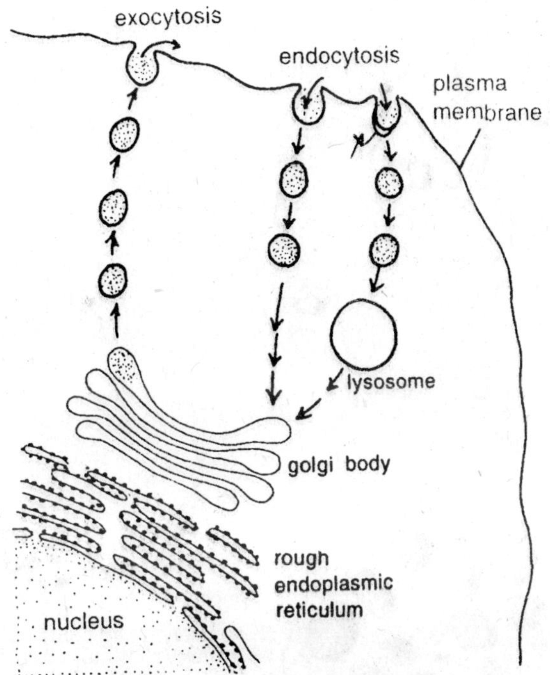

Fig. 3.4. Role of Golgi complex in the recycling and reuse of plasma membrane components.

produce mucigen. This secretory material contains a large proportion of carbohydrate. The Golgi complex is found just above the nucleus. Towards the free surface of the cell are gradually enlarging mucigen granules. The proximal cisternae of the Golgi complex do not show any swelling, but at some distance across the stack the distal cisternae are quite suddenly converted into mucigen granules. The distal cisternae continually convert into mucigen granules every 2-4 minutes. New proximal cisternae are formed in compensation.

Role in Secretion

Golgi complex is considered to play some role in the secretory functions of a cell. But the question is this that they are secreting or synthesizing some substances themselves or they are simply a store house in which the secretory products which are secreted some where else in the cell, is simply stored and concentrated.

From the studies of *Palade et. al.* 1962 this secretory cycle is now well-defined and includes four steps in case of pancreatic acinar cells and they are:

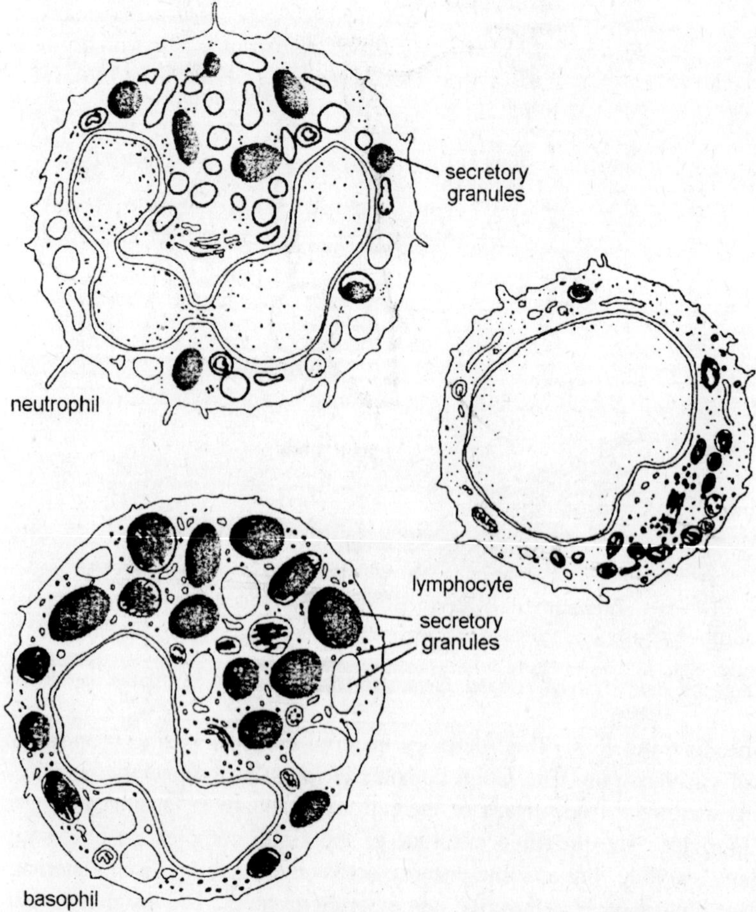

Fig. 3.5. Secretory granules in some leucocytes.

(i) The incorporation of amino acids into protein at the surface of rough endoplasmic reticulum.

(ii) Transfer of these nascent secretory proteins into the cisternae of rough endoplasmic reticulum.

(iii) The intracellular transport of these proteins to the Golgi complex.

(iv) The migration of *zymogen granules* towards the apex of the cell where they discharged into lumens.

Role of Golgi Body in Oogenesis

Srivastava (1965) has given a brief review on the Golgi complex during oogenesis. According to *Afzelius* (1956), the Golgi complex of a sea-urchin egg, as seen under electron microscope, consists of stacks of lamellae forming walls of flat pouches, which may occasionally be swollen. There is some indications of transverse divisions of these bodies. *Sotelo* (1959) and *Sotelo* and *Porter* (1959) have described the Golgi complex in rat-ovum as seen under the electron microscope and found juxtra nuclear localization of this organelle in early oocytes. In the next stage, these resolve into fragments and in the third stage, these move towards the cortex. In all these cases, their structures remain to be of closely packed arrays of slender, double profiles (flattened sacs) and spherical vesicles. In the early oocytes the complex is compactly organized. In later stages, discrete bundles of profiles, surrounded by small vesicles are found scattered in the cortical zone. In the early oocytes, the Golgi complex and centrosome are closely associated.

Absorption of Compounds

Hirsch et. al. have discovered that when iron sugar is fed to an animal, iron becomes absorbed on Golgi complex (*Kedrowsky*) *Van Teel* has shown that Golgi systems also absorb compounds of copper and gold. *Kedrowsky* has shown that Golgi complex of *Opalina* can absorb bismutose (compound of albumin and bismuth) and protargol (compound of albumin and silver). Thus, *Kirkman* and *Severinghaus* state that Golgi complex acts as a condensation membrane for the concentration of products into droplets or granules.

Plant Cell Wall Formation

The cell walls of plants are made up of fibrils which predominantly contain polysaccharides, along with some lipids and proteins. During cytokinesis a *cell plate* is formed between the two daughter nuclei, and has around it a membrane which later becomes the plasma membrane of the daughter cells. There is clear evidence that the polysaccharides are formed in the Golgi complex and transferred to the new cell wall which is laid down while the cells are still growing.

Substances like pectins and hemicelluloses, which form the matrix of the cell plate separating the plasma membranes, are also contributed by the Golgi complex.

Formation of Intracellular Crystals

In the marine isopod, *Limnoria ligmorum*, which is a burrowing form there are present midglands whose cells consist of crystals. These range up to 30Å in length and 15Å thick. It has been proved that these crystals are formed by Golgi complex and are known to contain protein and iron. They are without enclosing membrane and usually spheroidal in shape. They are concerned with the secretory activity.

Milk Protein Droplet Formation

In the lactating mammary gland of mice are produced protein droplets which are related with Golgi complex. These droplets usually open on to the cell surface by the fusion of their enclosing membrane with the plasma membrane.

Formation of Lysosomes and Vacuoles

Primary lysosomes are formed from the Golgi membranes the same way as the secretory vesicles. There is good evidence that dictyosomes accumulate hydrolytic enzymes in their more mature regions. Some vacuoles in plant cells have been found to certain small amounts of hydrolytic enzymes and these are presumed to have been derived from Golgi complex.

Pigment Formation

In many mammalian tumour and cancer cells the Golgi complex has been described as the site of origin of pigment granules (melanin).

Regulation of Fluid Balance

A homology has been suggested between the Golgi complex and the contractile vacuole of lower Metazoa are Protozoa. The contractile vacuole expel surplus water from the cell. In certain Protozoa the Golgi complex is also concerned with regulation of fluid balance.

Origin of Golgi Complex

The different sources have been proposed from which new Golgi complex may arise:

From Endoplasmic Reticulum

Essner and *Novikoff* (1962) and *Beams* and *Kessel* (1968) have proposed that the Golgi cisternae arise from the ER. The rough endoplasmic reticulum after synthesizing specific proteins loses ribosomes and changes into smooth ER. Small transitory vesicles pinch off from smooth ER. These migrate to dictyosome. On reaching the forming face of dictysome these fuse to form new cisternae and thus contribute to its growth. By the fusion of these vesicles new cisternae are formed continuously on the forming face and on the maturing face the old cisternae break down into secretory vesicles. Thus Golgi exhibit a phenomenon of 'membranous flow.'

From Nuclear Membrane

Bouch (1965) described the origin of Golgi from outer membrane of nuclear envelope in brown algae. Vesicles are pinched off from outer nuclear membrane which fuse to form cisternae on the forming

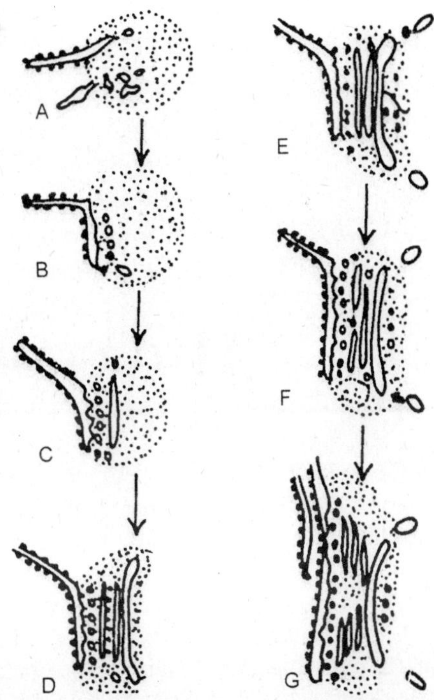

Fig. 3.6. Origin of dictyosome from endoplasmic reticulum from A to D and division of dictyosome—E, F, G.

face of dictyosome. Presence of zones of exclusion in relation with smooth ER or nuclear membrane, the occurence of zones of exclusion in dormant seeds of higher plants and the formation of dictyosome from these zones in germinating seeds provide evidence in support of the above two theories about the origin of dictyosome.

By the Division of Pre-existing Dictyosome

It has been observed that during cell division in both plants and animals, the number of dictyosomes increases and the number of dictyosomes in each daughter cell just after division is almost equal to the number in the parent cell prior to division. From this and other direct observations on the dividing cells it has been presumed that dictyosomes also divide during cell division.

4

MITOCHONDRIA

Kolliker (1880) was the first who observed the granules (mitochondria) in muscle cell of insects. *Flemming* (1882) named the mitochondria as *fila*. *Altmann* in 1894 observed them and they were called *Altmann's granules bioplasts*. The term *mitochondria* was applied by *Benda* (1897-98) to these granules which were described as *cytomicrosomes* by *Velette St. George*. *Benda* Stained mitochondria with alizarin and crystal violet. *Kingsbury* (1912) related them to cellular respiration and *Warburg* (1913) observed the presence of respiratory enzymes. In 1934 *Bensley* and *Horr* isolated mitochondria from liver cells and *Porter* and *Palade* described their electron microscopic structure.

Distribution

Ordinarily mitochondria are distributed in the cytoplasm. They may, however, be localized in certain regions. In proximal convoluted tubules of the kidney they are found in the basal region of the cell, opposite the renal capillaries. In skeletal muscles they lie between the myofibrils. In insect flight muscle several large mitochondria are in contact with each fibril. In cardiac muscle the mitochondria are situated in clefts between the myofibrils, numerous lipid droplets are associated with the mitochondria. In many sperms the mitochondria fuse into one or two structures which lie in the middle piece of the tail, surrounding the axil filament. In columnar or prismatic cells they are oriented parallel to the long axis of the cell. In leucocytes they are radially arranged.

Orientation

Mitochondria may have a more or less definite orientation. For example in cylindrical cells they are generally oriented in the basal-apical direction, parallel to the main axis. In leukocytes, mitochondria are arranged radially with respect to the centrioles. It has been suggested that these orientation depend upon the direction of the diffusion currents within cells and are related to the submicroscopic organization of the cytoplasmic matrix and vaculolar system.

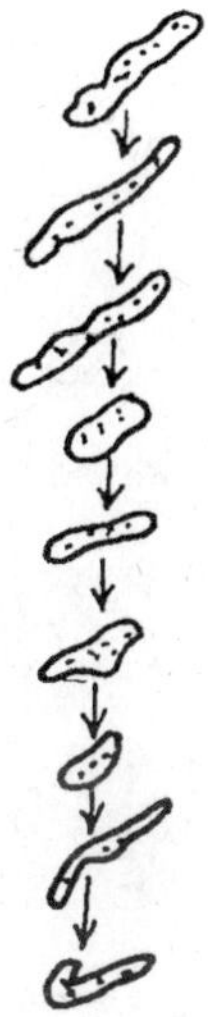

Fig. 4.1. Rapid changes of shape in living cells.

Plasticity of Mitochondria in a Cell

Lewis and *Lewis* (1914-15) concluded that the mitochondria are extremely variable bodies, which are continually moving and changing shape in cytoplasm. There are no definite types of mitochondria, as any one type may change into another. They appear to arise in cytoplasm and to be used up by cellular activity. The shape may change fifteen to twenty times in ten minutes; it can be changed by heat, hypertonic and hypotonic media or by acids, fat solvents, potassium permaganate and osmotic changes. *Frederic* (1958), *Littre* (1954), *Tobioka* and *Biesels* (1956) studied the effect of large number of chemical and physical agents on mitochondrial behaviour. Some material, such as detergents, show some effect *in vivo* as on mitochondria isolated from homogenates.

MORPHOLOGY

Shape

The shape is variable but is characteristic for a cell or tissue type, this too is dependable upon environment or physiological conditions In general they are *filamentous* or *granular*. They may swell out one end to become *club-shaped* or hollow out at one end to assume a shape of *tennis-racket*. They may become *vesicular* by the appearance of central clear zone. *Rod-shaped* mitochondria are also observable.

Size

The size of mitochondria also varies. In majority of the cells, width is relatively constant, about 0.5μ, but the length varies and, sometimes, reaches a maximum of 7μ. The size of cell also depends on the functional stage of the cell. Very thin mitochondria, about 0.2μ, or thick rods 2μ are also seen. The size and shape of the fixed mitochondria are determined by the osmotic pressure and pH of the fixative. In acid, pH mitochondria are fragmented and become vesicular.

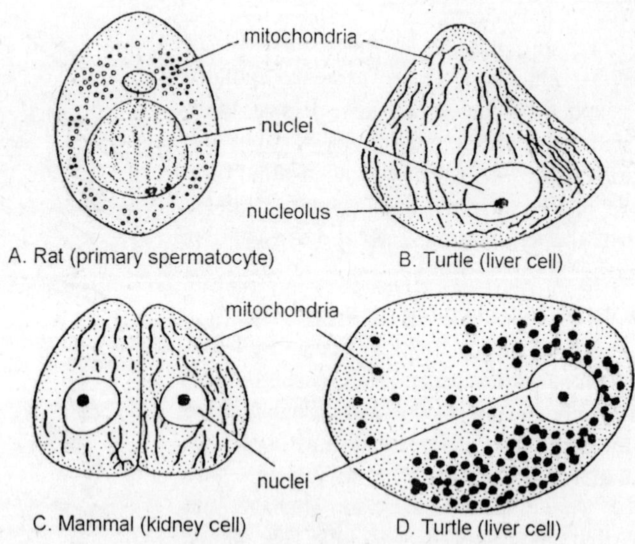

Fig. 4.2. Mitochondria of different types of animal cells.

Mitochondria, in the rat-liver, are usually 3.3μ in length; in mammalian exocrine pancreas, they are about 10μ in length and in oocytes of Amphibia, they are approximately of the length of 20 to 40μ.

Number

Mitochondria are found in the cytoplasm of all aerobically respiring cells with the exception of bacteria in which the respiratory enzymes are located in the plasma membrane. The mitochondria content of a cell is difficult to determine, but in general, it varies with the cell type and functional stage. It is estimated that in liver mitochondria constitute 30 to 35 percent of the total protein content of the cell and in kidney, 20 percent. In lymphoid tissue the value is much lower. In mouse liver homogenates there are about 8.7×10^{10} mitochondria per gram of fresh tissue. A normal liver cell contains about 100 to 1600 mitochondria, but this number diminishes during regeneration and also in cancerous tissue. This last observation may be related to decreased oxidation that accompanies the increase to anaerobic glycolysis in cancer. Another interesting finding is that there is an increase in the number of mitochondria in the muscle after repeated administration of the thyroid hormone, *thyroxin*. An increased number of mitochondria has also been found in human hyperthyroidism.

Thus cell with high metabolic activity have a high number of mitochondria, while those with low metabolic activity have a lower number. Large sea urchin eggs have 13,000-14,000, while renal tubules have 300-400. In the sperm there are as few as 20-24 mitochondria while in some oocytes there are about 300,000. In the protozoan *Chaos chaos* there are about 500,000 mitochondria. Some algal cells contain only one mitochondria.

Structure of Mitochondria

A typical mitochondria in sausage shaped with an average diameter of about 0.5m. When it is properly fixed in osmimum containing fluid and studied under electron microscope which reveals that there is hardly any difference between plant and animal mitochondria. In both the cases the mitochondria is bounded by two membranes, the *outer membrane* and the *inner membrane*. The space between the two membranes is called the *outer chamber* or *inter-membranous space*. It is filled with a watery fluid, and is 40-70Å in width. The space bounded by the inner chamber is called the *inner chamber* or *inner membrane space*. The inner membrane space is filled with a *matrix* which contains dense granules (300-500Å), ribosomes and mitochondrial DNA. The granules consist of insoluble inorganic salts and are believed to be the binding sites of divalent ions like Mg^{++} and Ca^{++}.

In some cases, they apparently contain polymers of sugars. The side of the inner membrane facing the matrix side is called the *M-side*, while the side facing the outer chamber is called the *C-side*. Two to six circular DNA molecules have been identified with mitochondria. These ring may either be in the open or in the twisted configuration. They may be present free in the matrix or may be

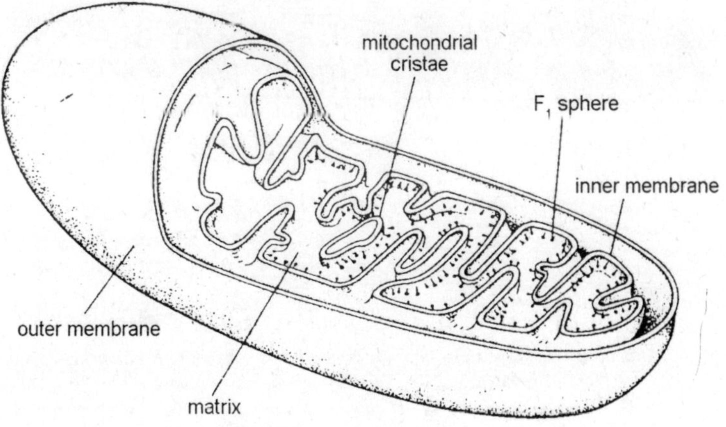

Fig. 4.3. Diagram showing the fine structure of a mitochondrion.

attached to the membrane. The enzymes of the Krebs' cycle are located in the matrix.

The inner membrane is thrown up into a series of folds, called *cristae mitochondriales*, which project into the inner chamber. The cavity of the cristae is called the *intercristae space*, and is continuous with the inner-membrane space.

The space and arrangement of crests is variable and may be of the following types:

1. Parallel to the long axis of mitochondria as in the neurons and the striated muscles cells.
2. Concentrically arranged as in the matrix of certain spermatids.
3. Interlaced to form villi as in *Amoeba*.

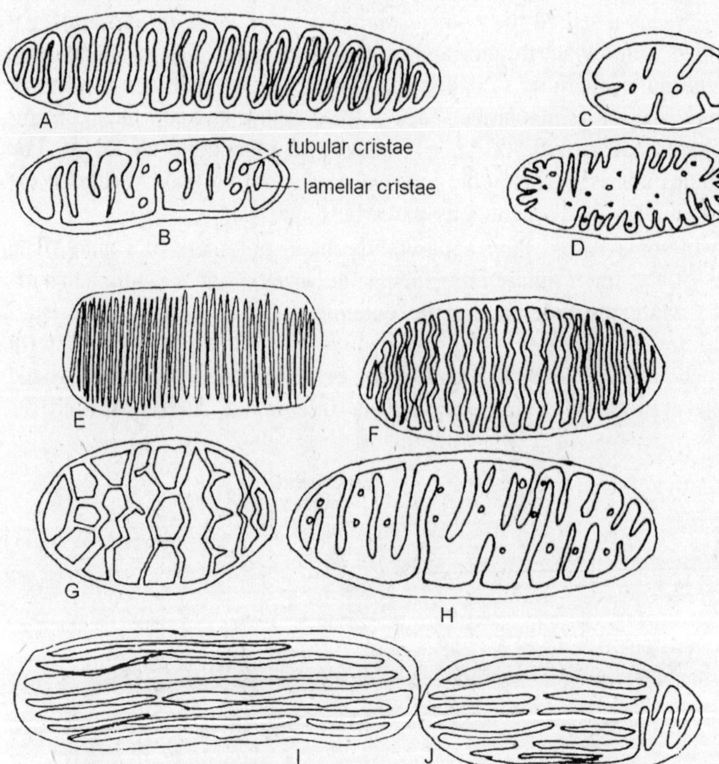

Fig. 4.4. Structural variations in mitochondrial cristae. A—Tubular, B—Lamellar cristae, C—Cristae in salamander, D—Plate like cristae, E—Parallel cristae, F—Cristae with sharp ongulation, G—Honey corubs cristae, H—Usual cristae, I—Longitudinal cristae, J—Transverse and longitudinal cristae.

4. Cristae in the form of vesicles which form a network of interconnected chambers as in the cells of parathyroid gland and W.B.C. of man.
5. Arranged in a tubular fashion but perpendicular to mitochondrial axis as in the cells of adrenal gland.
6. Haphazardly distributed as in the cells of kidney of insects and hepatic cells.
7. Cristae externally small and irregular as in the interstitial cell of *Opossum*.
8. Rarely the mitochondrial wall is smooth with no cristae. The number and size of cristae in a mitochondrion directly affects its efficiency. The greater and larger are the cristae, the faster is the speed of oxidation reaction.
9. Perpendicular to the long axis of mitochondria.

Mitochondrial particles

According to earlier description, the outer surface of the outer membrane and the inner surface of the inner membrane were supposed to be covered with thousands of small particles. Those on the outer membrane were described as being stalk less and were called the *subunits* of *Parson*. There may be as may be as many as 10,000 to 100,000 particles per mitochondrion. However recent studies have shown that stalkless particles are absent. The stalked inner membrane particles *were called the subunits of Fernander-Moran*, *Elementary particles*, F_1 *particles* or the *oxiosomes* or *ETP* or *electron transport particles*. These particles are about 84Å in diameter and are regularly spaced at intervals of 10mm on the inner membrane. There may be as many as 10^4 to 10^5 elementary particles per mitochondrion.

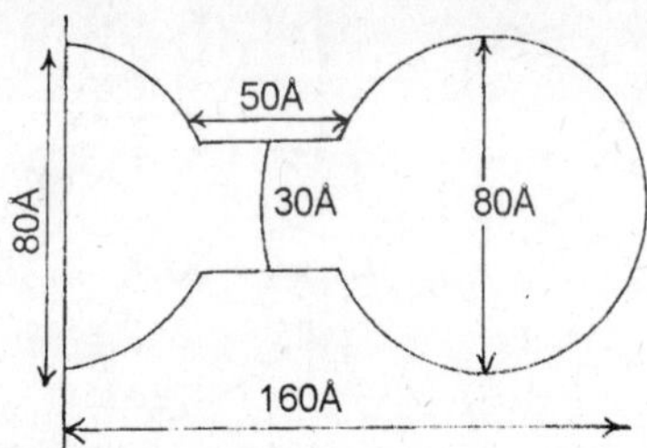

Fig. 4.5. Structure and measurements of the mitochondrial particles.

Isolation of Mitochondria

Mitochondria can be isolated from the cell in the living form for their physiological studies. The cell first treated with deoxycholate for their break down. Then they are passed in sucrose solution. The homogenate should be centrifuged for 10 minutes at the speed of 6000 X g. From this homogenate upper substance is centrifuged at the speed of 8500 X g for 10 minutes. After this centrifugation, the upper microsomal fraction is discarded while the lower fraction consists of mitochondria and other particles like lysosomes. The mitochondrial fraction then centrifuged at the speed of 10,000 X g upto 3 hrs. The upper part of this centrifuged material have mitochondria and lower part of lysosomes.

Respirating Chain Complexes

Green et. al. have recognized five main complexes which, if mixed in correct ratios, can reconstitute to form ETC. These complexes are:

Complex I (NADH-Q-reductase)

This is largest complex, with a molecular weight about 500,000 and structure consisting of 15 subunits. It contains as prosthetic groups *flavin mononucleotide* (FMN) and six iron-sulphur centers. The NADH reaction site lies at M side of mitochondrion. Contract between NADH

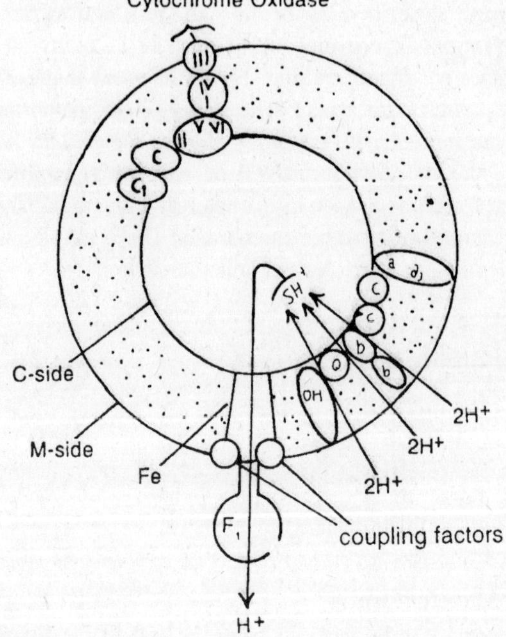

Fig. 4.6. Inner membrane of mitochondrion showing distribution of respiratory chain complex.

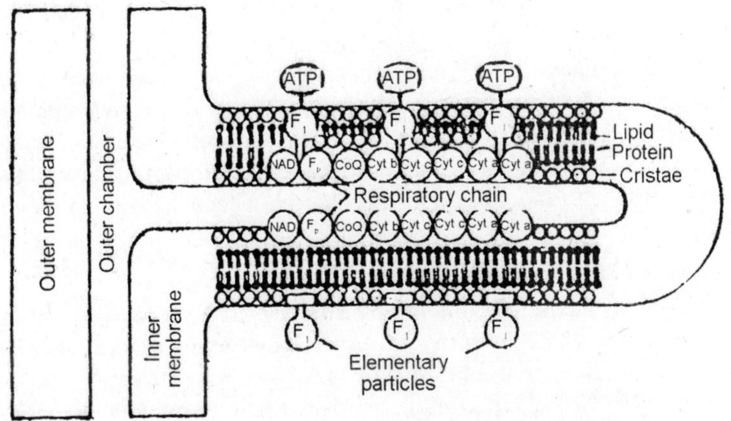

Fig. 4.7. The arrangement of the respiratory chain in cristae.

and CoQ is apparently made in the middle of the membrane. Complex I spans the inner mitochondrial membrane made and is able to translocate the protons across it, from M side to C side of mitochondrion.

Complex II (Succinate-Q-reductase)

This complex is made up of two polypeptides with a molecular weight of 97000. It contains *flavin adenine dimucleotide* (FAD) and three iron-sulphur centers (Fe-SS_1, Fe-SS_2 and Fe-SS_3). The succinate binding site occurs on the M side. The three Fe-S centers are on the M side and there is a close interaction between Fe-S centre 3 and CoQ. In contrast to complex I, succinate-Q-reductase apparently is unable to translocate protons across the membrane.

Complex III (Q H_2-cytochrome-c-reductase)

This complex contains a number of subunits with a molecular weight 280,000. It contains cytochromes b, cytochrome C_1 and iron sulphur protein. There are two types of cytochrome b, these are bT (transducing cytochrome b) and b_k (Keilin-type cytochrome b). The heme containing part of cytochrome C_1, which transfer the electrons to cytochrome-c, is located on the C side of mitochondria.

Complex IV (Cytochrome-C-exidase)

Cytochrome oxidase is a large complex consisting of several polypeptides.

It has two cytochromes, a and a_3, and two copper atoms. The molecular weight is about 200,000. The a and a_3 however, never been separated and, therefore, both must be considered to be parts of the same complex.

Complex IV is thought to transverse the mitochondrial membrane, protruding on both surface.

In experiments with yeast mitochondria it has been shown that cytochrome-c-oxidase is made of seven subunits. The seven subunits are arranged the membrane in a functional sequence, being in contact with cytochrome c on the C-side. The electrons then pass to cytochrome a then to Cu^{++} and finally to cytochrome a_3 and oxygen at the M side.

Complex V (ATPase complex)

Towards the M side, the inner membrane contains rounded stalked particles known as *F_1 particles* or *Fernandez-Moran particles*. Each F_1 particles is made up of a head, stalk and base. It has been shown that complex V is identicle with F_1 particles. These are four compling factors on the M side designated as F_1, F_2, F_5 or OSCP (oligomycin-sensitive confering protein) and F_6. The headpiece, coupling factor F_1, is the ATPase proper. The F_1 contains five types of subunits, α with a molecular weight 53,000 daltons, β with a molecular weight 50,000 daltons, γ with a molecular weight 33,000 daltons, δ with a molecular weight 17,000 daltons and ε with a molecular weight 7,000 daltons. In addition to these subunits there is ATPase inhibitor (I) with a molecular weight 10,000 dalton. This inhibitor can be removed while treating with trypsin.

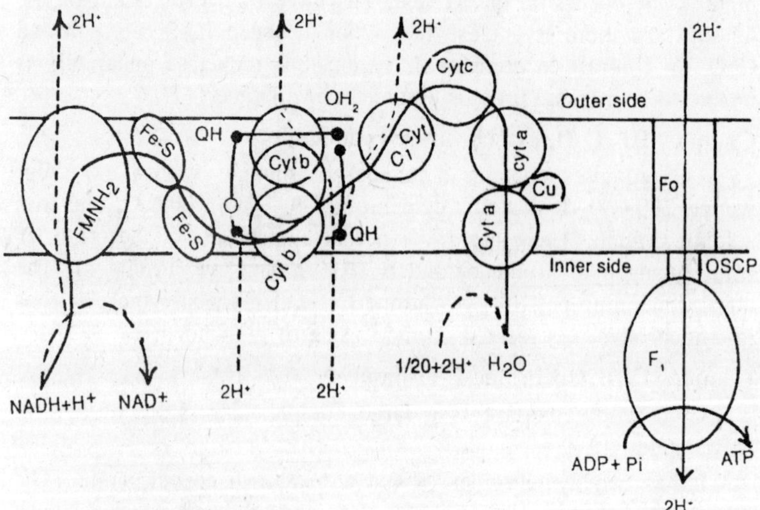

Fig. 4.8. Current view of the location of the enzymes and components of the respiratory chain in the inner membrane of the mitochondrion.

The stalk, made up of protein, connects the head-piece to the base. This portion corresponds to the OSCP (oligomycin—sensitive conferring protein) and F_6. These are needed to bind F_1 to the membrane. Treatment with ammonia releases OSCP. While treatment with silicotungustate removes F_6.

The basepiece lies within the inner mitochondrial membrane. It contains the proton translocating mechanism. This corresponds to Fo in the figure 4.8.

Biochemistry of Mitochondria

Lindberg and *Ernster* (1954) have given the data of chemical composition of mitochondria as follows: proteins 70 to 75%, Lipids 25-30%, and RNA 5% of the dry weight. But the recent biochemical analaysis shows the following components:

Proteins

The proteins are the main constituent which is insoluble in water. The outer limiting membrane of mitochondria contains less than 10 percent of the total protein. There are about 14 different proteins having molecular weight from 12,000 to 22,000. The inner membrane contains about 60% protein having molecular weight varying from 10,000 to 90,000. The protein composition of mitochondrial membranes is not fully known.

Lipid

The lipid forms about 1/5th of the weight of the membranes. It is present almost entirely in the form of the molecules known *phospholipid*. It has been reported by *Meluick* and *Packer* in 1971 that the outer membrane fraction has a 40% lipid contents as compared to 20% in the inner membrane.

Enzymes

About 70 enzymes and 12 co enzymes have been recognized in the mitochondria. Enzymes lie in a non-aqucors membraneous region as solid arrays, with perhaps as many as 5000 to 20,000 such assemblages in a single liver or heart mitochondrion.

Mitochondrial DNA

Recently the DNA is also reported from mitochondria. The mitochondrial DNA is double stranded like the nuclear DNA. Each mitochondrion may contain one or more DNA molecules depending on its size, if the mitochondrion is larger than that may have more DNA molecules, having a circular shape.

Table 4.1. Localization of enzymes, obtained in fractionation studies.

Mitochondrial fraction	*Enzymes located*
1. Outer membrane.	Monoamine oxidase, "Rotenone-insensitive" NADH-cytochrome – C – reductase, Kynurenine hydroxyl-ase, Fatty acid CoA ligase, Glycerol-phosphate acyl transferase, Nucleoside diphosphokinase.
2. Inter-membrane space.	Adenylate kinase, Nucleoside diphosphokinase, Nucleoside mono-phosphokinase,
3. Inner membrane	Respiratory chain enzymes, β-Hydroxybutyrate, dehydrogenase, Ferrochelatase, Carnitine palmityl-transferase, Fatty acid oxidation system, Xylitol dehydrogenase.
4. Matrix	Malate, isocitrate and glutamate dehydrogenase, Fumarse, Aconitase, Citrate synthetase, Ornithine-Carbonyl transferase, Fatty acid oxidation system, Pyruvate carboxylase,

Mitochondrial DNA differs from nuclear DNA in several respects. The GC content is higher in mitochondrial DNA, and consequently the buoyant density is also higher. Another difference is the higher denaturation temperature of mitochondrial DNA and facility with which it denatures. The amount of genetic information carried by mitochondrial DNA is not sufficient to provide specifications for all the proteins and enzymes present in this organoid. The most likely possibility is that mitochondrial DNA codes for some structural proteins.

Yeast mitochondria have been shown to contain DNA polymerase and more recently *Kalf* 1968 has succeeded in isolating the enzymes from rat liver mitochondria. Mitochondrial DNA polymerase appears to be involved in DNA replication rather than repair and possesses properties which are different from those to the nuclear enzymes. These include a differing requirement for metal ion. *Yeast* mitochondrial DNA polymerase appears to be smaller than its nuclear counter part, and is active at different stages of the cell cycle. Visual evidence showing what appear to be rat liver mitochondrial DNA in the process

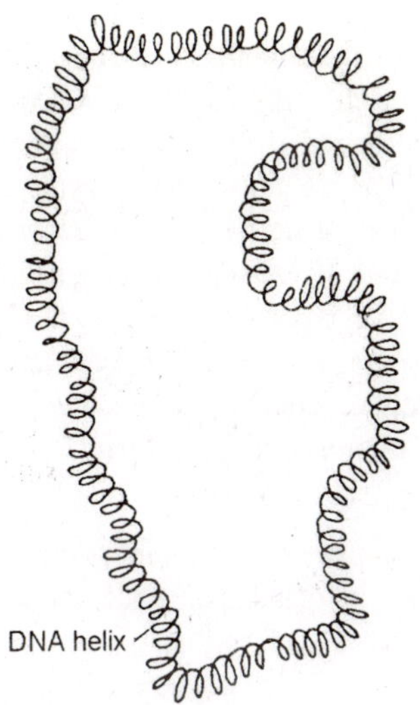

Fig. 4.9. Mitochondrial DNA.

of replication has been presented by *Kirschner, Wolsten Holme* and *Gross* (1968). Mitochondrial DNA does not appear to have histones associated with it as has nuclear DNA of higher organisms. In this respect mitochondrial DNA resembles with the bacterial DNA.

Mitochondrial RNA

South and *Mehlear* (1968) suggested that the amount of mt-RNA is about 10 to 20 times that of mt-DNA. All sorts of RNA have been identified in mitochondria. The present evidences point out conclusively that mitochondria contain complete set of *t*-RNA (*Wintersberger* and *Tuppy,* 1965), aminoacyl RNA synthetases (*Barnett, Braun and Epler,* 1967), as well as ribosomal RNA (*Rogers, Preston, Titchener* and *Linnane,* 1967). All these components differ from their respective counterparts in the groundplasm. The presence of m-RNA, transcribed from mitochondrial DNA is still uncertain. However, there are authorties who suggests its presence. The ribosomal RNA's are coded for by mitochondrial DNA and are thus apparently synthesized within the mitochondria by a mitochondrial DNA-dependent RNA polymerase system (*Wintersberger,* 1964).

Mitochondrial Ribosomes

Mitochondria appear to contain ribosomes which are smaller in diameter than cytoplasmic ribosomes (*Swift,* 1965) and *Yeast* mitochondria contain RNA species of 23^s and 16^s (*WinterBerger* 1966) which would correspond to a 70^s ribosome of bacterial type rather than the 80S ribosome of the cytoplasm. Ribosome like particles with sedimentation values of 81S and 55S have also been reported, and the extent of degradation suffered by the particles during isolation is not yet clear. Polysome like aggregation of ribosomes have been observed in sections of yeast mitochondria by *Vignais, Huet* and *Andre* in 1969. High molecular weight and RNA species associated with mitochondria which differ in sedimentation value from cytoplasmic ribosomal RNA have been reported in *Yeast*, *Neurospora* and *He-La cells*. Mitochondrial ribosomes require a higher concentration of Mg^{++} ions to maintain their integrity than do cytoplasmic ribosomes.

Protein synthesis

In general, mitochondria can code and synthesize protein, but the DNA present therein is unsufficient to code for all the proteins. It is suggested that mitochondria can synthesize the proteins of structural nature (cytochrome oxidase), but much of the proteins if not all, of the soluble proteins of the matrix as well as the proteins of the outer membrane and a number of proteins located in the cristae (*Borst,* 1972) are under the control of nuclear DNA. Of the proteins coded by nuclear DNA, it is generally agreed that the m-RNA derived from nucleus are translated in the cytoplasm, and the resulting proteins are then transported into the mitochondria. How then protein enter the mitochondria? Two methods have been proposed:

1. The precursors enter the mitochondrion and inside are changed into the end products, thereby affecting a undirectional flow of material into the mitochondrion.
2. There is the synthesis of lipoprotein vesicles which they merge and combine with the growing mitochondrion.

FUNCTIONS

Role of Mitochondria in Yolk Formation

There has been a good number of investigations, whose account reveals that, mitochondria help in the formation of yolk in a developing ovum. The first study in this field was made by *Loyez* (1911) and latest probably by *M.D.L. Srivastava* (1965), with the help of light microscope. The evidence adduced depends upon the topographical, and size relationship, and staining reactions of mitochondria and early protein yolk.

In the modern cytology with the investigation of electron microscope a new era has started and the studies of yolk formation do not remain away from electron microscope. With the help of electron microscope *Farvard* and *Carasso* (1958) come to the conclusion that mitochondrial transformed into yolk granules in the egg of *Planorbis coneus*.

The main structural changes which they observed in the mitochondria are as follows:

(i) The cristae become disorganised in a few membranes, remaining concentric to the other membrane before dropping completely.

(ii) In the matrix appeared a few minute granules which have scattered first, but become aggregated eventually in masses in regular pattern.

During Cell division and Spermiogenesis

Early cytologists, *Benda, Dulberg* and *Meves* were of opinion that mitochondria also divide equally during the cytoplasmic division and perhaps playing a part in inheritance.

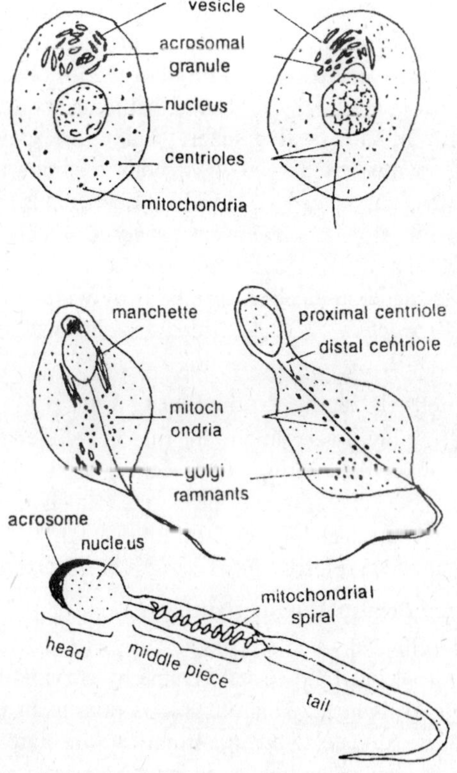

Fig. 4.10. Role of mitochondria during spermatogenesis.

Wilson (1928) commented that not the slightest proof has been produced of a fusion between the paternal and maternal chrondrisomes. *Frederic* (1958) has briefly summarized various changes in mitochondria during cell division. The first phase shows a decrease in the total volume of mitochondrial material; gradually ceases its movements, pronounced thinning, fragmentation into small spheres, loss of optical density and finally assimilation into the cytoplasm. In the second phase, when the cell divides into two, the modified mitochondria are separated passively, into the daughter cells. In the third phase, the modified mitochondria are reconstituted by addition of elements assimilated in the cytoplasm. *Wilson* found in *Opisthacanthus*, during spermatogenesis, the number of mitochondria gradually reduces. *Pollister* (1930) describes in *Gerris* that mitochondria arranged themselves into a well-defined ring, but without fusion. Modern microscopic studies provide firm conclusions regarding mitochondrial division during mitosis. *Pyne* (1952) noted in fowl adrenal cortex the mitochondria frequently appeared as pairs. This suggested that division, rather than fusion, was occurring. In the transformation of spermatids into spermatozoa many mitochondrial changes are observed. *Franzen* (1956) observed in those sperms, which are shed directly in water, that mitochondria are present in form generally of four or five spheres below the sperm head, and in the case of sperms discharged in viscous medium these spheres transform into two enlongated ribbion like filamentous mitochondria. Sometimes these develop into '*nebenkern spheres*' which may elongate and twist around the axial filament to form the mitochondrial sheath. *Yasuzumi* (1958) found an electron opaque body within 'nebenkern' of spermatids of *Drosophila;* it is described as indistinguishable from a lipid droplet.'

Role of Mitochondria in the Production of Energy

Mitochondria play a very important role in cellular respiration or the production of energy. Energy is produced inside the cell, partly outside the mitochondria and mainly inside the mitochondria. The ATP molecules that are produced in the non-mitochondrial cytoplasm are generated by a process referred to as *anaerobic respiration*.

Role of Mitochondria in Haeme Synthesis

In rat liver cells and avian red blood corpuscles d-amino levulinate is synthesized for succinyl Co-A and glycine by enzymatic action of d-*amino levulinate synthatase*. This enzyme is present in mitochondrial fraction. d-aminolevulinate is an important intermediate in prophyrin synthesis. This mitochondria help in haem synthesis.

Role in Gluconeogenesis

The gluconeogenesis is the conversion of non-carbohydrates to glucose from pyruvic acid. It is well-known that pyruvic acid is converted into oxaloacetic acid in the presence of pyruvic acid carboxylase. This intermediate may escape mitochondria and become converted into phosphoenol pyruvic acid by phosphoenol pyruvate corboxy

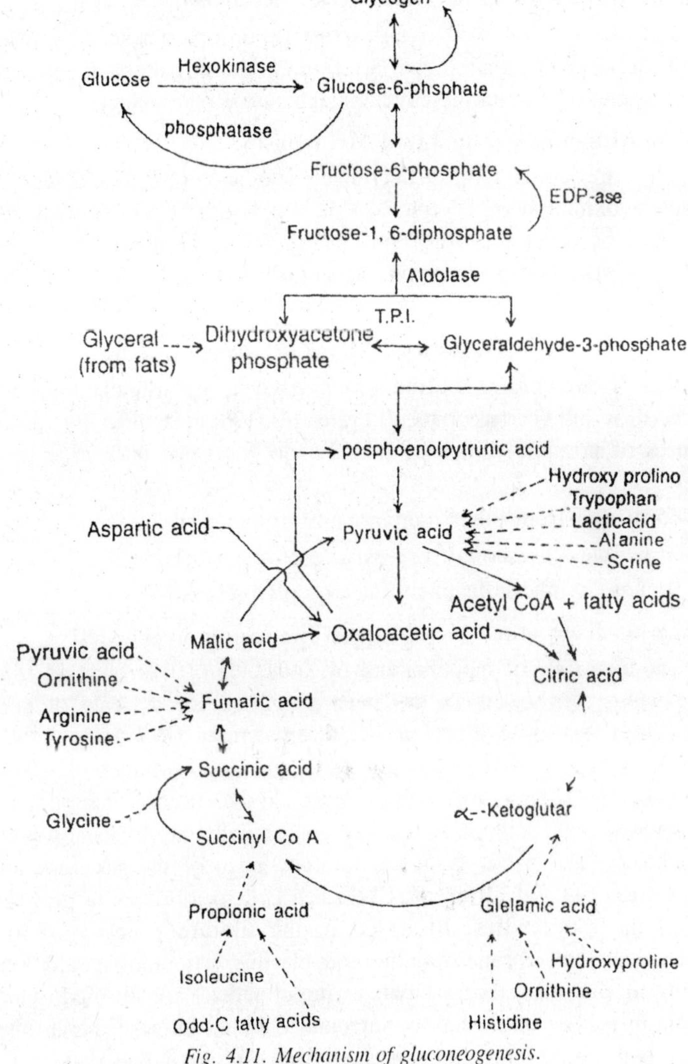

Fig. 4.11. Mechanism of gluconeogenesis.

kinase. Phosphoenol pyruvic acid occupies a place in Embden-Mryerhoff pathway or glycolytic pathway from where the pathway is reversible upto glucose. Glucogenic amino acids, lactic acid, glycerol and in some cases propionate can, after suitable modification, be fed at one or the other point in Krebs' cycle. Oxaloacetic acid and malic can come out of mitochondria to be finally converted into glucose.

Role of Mitochondria in Amino Acid Metabolism

The enzymes for oxidative deamination of amino acids are present in mitochondria. These are glutamate dehydrogenase, proline dehydrogenase, d-amino levulinate synthetase etc.

Role of Mitochondria in Lipid Metabolism

They are able to oxidize fatty acids. Oxidation of fattyacid requires complete oxidation of acetyl CoA in the Krebs' cycle so that free CoA may be generates. Reversal of fatty acid oxidation leads to fatty acid synthesis. During starvation, the mitochondria utilize fat to produce energy.

Origin of Mitochondria

Our understanding of the process where by mitochondria are produced, is still very incomplete. *Lehninger* (1964) classified the various theories of possible routes of mitochondrial genesis into three main groups:

1. Formation from other membranous structures in the cell.
2. Growth and division of pre-existing mitochondria.
3. *De novo* synthesis from submicroscopic precursors.

Formation from other Membranous Structures in the Cell

The formation of mitochondria by "pinching off" or budding from pre-existing cell structures has been suggested for a range of cells membranes including those of the plasmalemma (*Robertson* 1959), endoplasmic reticulum, nuclear envelope, and Golgi complex (*Novikoff,* 1961). But the support for such evidence, in the absence of supporting biochemical data, cannot be wholly conclusive. Part of the problem undoubtedly lies in our fragmentary knowledge of the structure and composition of, and differences between cell membranes in general. Indeed the similarities discussed in the literature between the mitochondrial membrane and the endoplasmic reticulum, could lend weight to the idea that probably mitochondria are formed when cytoplasm pushes into a cavity surrounded by an internal membrane, which they, pinches off and separates from the continuous system.

Growth and Division of Pre-existing Mitochondria

Electron microscopic evidence for mitochondrial division by fission, although plentiful, is difficult to assess the danger of producing artifacts is very real because of the harsh chemical and physical agents brought to bear on the test material during processing. Interpretation is not made easier by the ability of mitochondria of undergo extreme changes in shape *in vivo* which may or may not be associated with mitochondrial fission. There are numerous reports of mitochondria connected to each other by narrow bridges of membranes, especially in rapidly melopolizing tissue, and it is thought that such figures may represent mitochondria in an early stage of fission. By observing serial sections of rat liver *Stempak* (1967) was able to show that "dumb bell-shaped" mitochondria can be the sections of cupshaped bodies. Such bodies have also been observed in rapidly growing tissues of fern and may represent the beginning stages of division.

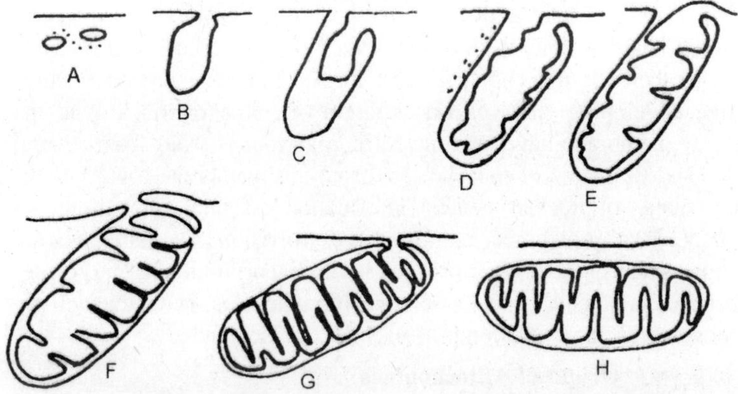

Fig. 4.12. A possible mechanism for the formation of mitochondria from the plasma membrane.

An early stage in mitocho-ndrial division may involve the separation of mitochondrial contents into two or more compartments. The presence of mitochondria with internal "partitions" has been formed in several cell types (*Tandler et. al.*, 1969) although the possibility that they are manifestations of mitochondrial fusion cannot easily be ruled out. *Lafontaine* and *Allard* (1964) have presented electron micrographs of rat liver mitochondria which exhibit what appear to be partitions dividing the inner membrane complex into two masses, the whole being surrounded by continuous outer membrane. *Tandler et. al.* (1969) has demonstrated the partitions of mitochondria in liver which was recovering from riboflavin deficiency.

De novo Synthesis

The possibility of *de-novo* synthesis of mitochondria arose with experiments in the early part of the century, when mitochondria containing larvae were seen to develop from sea urchin egg cytoplasm which had apparently been freed of mitochondria by centrifugation. Using the greater resolving power of the electron microscope, it was later shown, that mitochondria could not be dislodged by centrifugation of the egg. In the earlier experiments, mitochondria had probably been present in the "centripetal end" of the egg cell after all, and these mitochondria could have served as precursors in subsequent mitochondrial production.

In the above description, a number of views were described, regarding the generation of mitochondria in the cytoplasm of different types of cells. But it is perhaps unwise to group the evidence to suit one or other of a limited number of clear cut methods by which mitochondria could replicate. The actual situation is probably complex, and it may well be that different methods of replication take place in different tissues, and at different stages in development. One could imagine the early mitochondria being formed from membrane structures in the developing embryo concentration of mitochondria around the nuclear membrane have been noted in embryonic tissues from several phyla and formation of mitochondria from this membrane could involve the transfer of nuclear genetic information essential for subsequent mitochondrial growth and multiplication by division. The multiplication of mitochondria could then proceed by the incorporation of large pre-fabricated molecules and association of molecules, with division by fission, when the mitochondria reached a critical stage.

Prokaryotic Origin of Mitochondria

This fact that mitochondria can grow, divide and are capable of mutations support a long-held view that mitochondria originated with their host. Bacteria would have originated the mitochondria and blue green algae, the chloroplasts.

There are numerous homologies between mitochondria and bacteria. In bacteria the electron transport system is localized in the plasma membrane which can be compared with the inner membrane of mitochondria. Some bacteria even have membranous projections extending from the plasma membrane which are comparable to mitochondrial crests since both contain the respiratory chain. The inner membrane and matrix, it has been postulated may represent the original symbiont which may become enclosed within a membrane of cellular origin (ER). Further mitochondrial DNA is circular, it replicates and divides like that of bacteria. The ribosomes are also found which are, however, smaller than those of the bacteria. In mitochondria and bacteria, the protein synthesis is inhibited by chloramphenicol. From these similarities, one can easily conceive of mitochondria as being evolved from an ancient prokaryote, possessing all the attributes of an independent, probably aerobic organism. However, with the adaptation over a long period it became an essential and dependent symbiont, lost some of its identity to the cell, and conversely, the host cell lost some of its functions, deriving it now from the endosymbiont or mitochondrion. As a results, both became obligatory symbionts to each other.

This symbiont hypothesis for the origin of mitochondria and plastids has acheived wide popularity, but all biologists do not necessarily accept it. *Raff* and *Mahler* (1972) concluded that "while the symbiotic theory may be esthetically pleasing, it is not compelling." They presented a lot of evidences and proposed that mitochondria arose by inward blebbing from plasma membrane, by the acquisition some how of an outer membrane, and by the additional acquisition of a DNA genophore from the DNA of the protoeukaryote in which the evolution of mitochondria occurred. *Borst* (1972) proposed an *episome theory* and supposed that the DNA of mitochondrion left the "nuclear' DNA by a sort of amplification to become mapped within a membrane containing the respiratory chain.

5

LYSOSOMES

The concept of the lysosome originated from the development of cell fractionation techniques by which different subcellular components are isolated. By 1949 a class of particles having centrifugal properties some what intermediate between those of mitochondria and microsomes was isolated by *de Duve* and found to have a high content of acid phosphatase and other hydrolytic enzymes. Because of their enzymatic properties they were named lysosomes (Gr. *Lysis* = dissolution, *soma* = body). According to *Gohan* (1972) membrane bounded storage granules containing digestive enzymes are considered lysosomes of plant Alls. Hence *spherosomes*, *aleurone granules* and *vacuoles* of plant cells are supposed to have lysosome like functions.

Occurrence

With the exception of mammalian R.B.C., the lysosomes have been reported practically from all the animal cells. The presence of lysosomal particles have also be suspected (and in some cases established) in protista (protozoan, slime-moulds, fungi, algae and prokaryotic protista). In plant cells, considering the evidences as a whole, there now seems little doubt about their presence. Further, they have strong affinities with the lysosomes of animals and protista. In plants, further they should not be confused with spherosomes in function. According to *Pitt* lysosomes and spherosomes are two different organelles and the later are comparable to lipid droplets of animals. *Yatsu* and *Jack* (1972) have clearly shown that spherosomes are morphologically distinct organelles. *Gahm* (1973) reviewed the occurrence and histochemistry of plant lysosomes.

Morphology

Shape and Size

The shape and size of lysosome is variable. Morphologically they can be compared with *Amoeba* and white blood cells (W.B.C.). Due to their changing habit they cannot be accurately identified as the basis of their shape. Normally lysosomes vary in size from 0.4 to 0.8μ, but they may be as large as 5μ in mammalian kidney cells and are exceedingly large in phagocytosis.

Structure of Lysosomes

Like other cytoplasmic complexes, lysosomes are like round tiny bags filled with dense material and digestive enzymes. They consist of two parts (1) limiting membrane and (2) inner dense mass.

1. *Limiting membrane.* This membrane is single, unlike that of mitochondria and composed of lipoprotein. The chemical structure is homologous with the unit membrane of plasma-lemma consisting of bimolecular layer as suggested by *Robertson*.
2. *Inner dense mass.* This enclosed mass may be solid or of very dense contents. Some lysosomes have a very dense outer zone and less dense inner zone. Some others have cavities of vacuoles within the granular material. Usually they are supposed to possess denser contents than mitochondria.

They show their polymorphic nature and their contents vary with the stage of digestion, as they help in intracellular digestion.

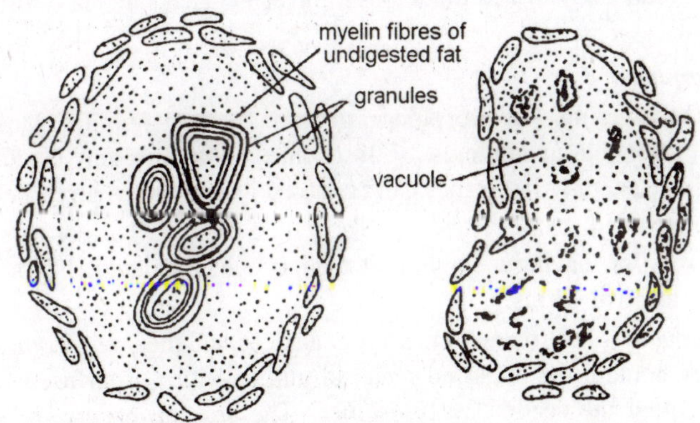

Fig. 5.1. Two types of lysosomes from the kidney cell of rat.

Permeability of lysosomal membrane

The lysosomal membrane is impermeable to substrate of the enzymes contained in the lysosome. Certain substances, called *labilizers*, cause instability of the lysosomal membrane, leading to release of enzymes from the lysosome. Other substances, called *stabilizers*, have a stabilizing action on the membrane. A list of some labilizers and stabilizers is given in table 5.1.

Table 5.1. Lysosomal labilizers and stabilizers.

Labilizers	*Stabilizers*
Vitamin A	Cholesterol
Vitamin B	Cortisone
Vitamin K	Cortisol
Vitamin E (high dose)	Vitamin E (low conc.)
Progesterone	Chloroquine
Testosterone	Phenothiazines
β-estradiol	Antithistamines
Ubiquinone	Heparin
Digitonin	
Polyene antibiotics	

A labilizer might increase the permeability of the lysosomal membrane to small solutes like source. The osmotic swelling which results might completely disrupt the membrane.

The limited permeability of lysosomal membrane explains why lysosomal hydrolases do not have a direct access to cellular components. This prevents uncontrolled digestion of the cell contents by the lysosomal enzymes.

Polymorphism

Lysosomes are polymorphic in nature. The polymorphic nature is due to variation in contents of lysosomes with different stages of digestion. Generally lysosomes can be traced in four forms given below:

1. *Primary lysosomes*. These are also called the *true*, *pure* or *original lysosomes*, having a single unit mem-brane containing enzymes in the inactive forms.
2. *Secondary lysosomes*. These are also called the *phagosomes* as they contain the engulfed material and enzymes. The fused mass is called the secondary lysosomes. The enzyme present in such lysosomes gradually digests the engulfed material.
3. *Residual or Post lysosomes*. Lysosomal membrane character-ized by the presence of undigested material like myelin figures in called residual body.

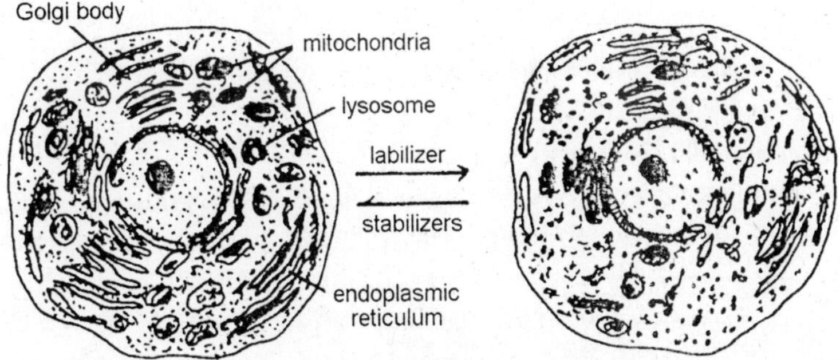

Fig. 5.2. Activities of the sucide bags in the cell.

4. *Autophagic Vacuoles.* The autophagic vacuoles are also known as *autophagosomes* or *cytolysosomes*. The autophagic vacuoles are formed when the cell feeds on its intracellular organelles such as the mitochondria and endoplasmic reticulum by the process of autophagy. In such caeses, the primary lysosomes are concentrated around the intracellular organelles and digest them ultimately. The autophagic vacuoles are formed in special pathological and

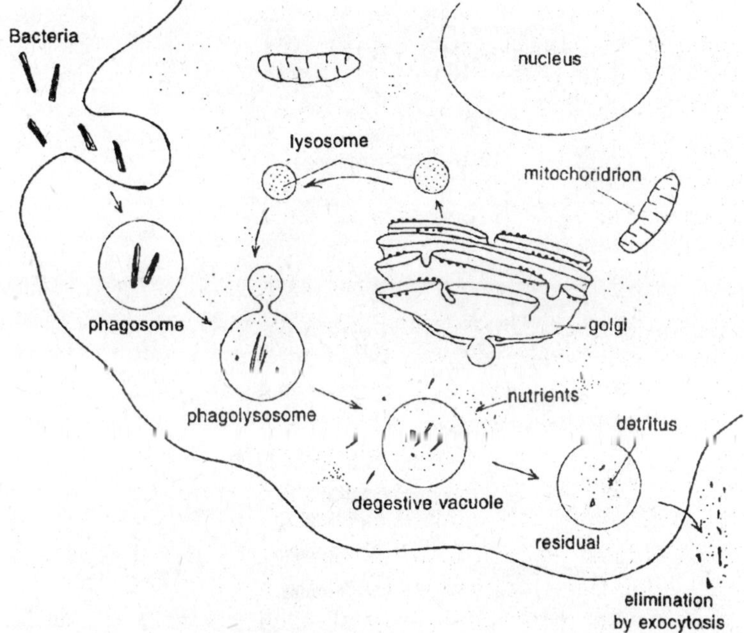

Fig. 5.3. Schematic representation of phagocytosis, showing ingestion, intracellular digestion and exocytosis.

physiological conditions. *C. de Duve* (1967) and *Allison* (1967) have observed that during starvation of the organisms many autophagic vacuoles developed in the liver cells which feed on the cellular components.

Excretion of Lysosomes

The phenomenon of centrifugation has played a great role in the study of cytoplasmic inclusion. The technique about the separation of

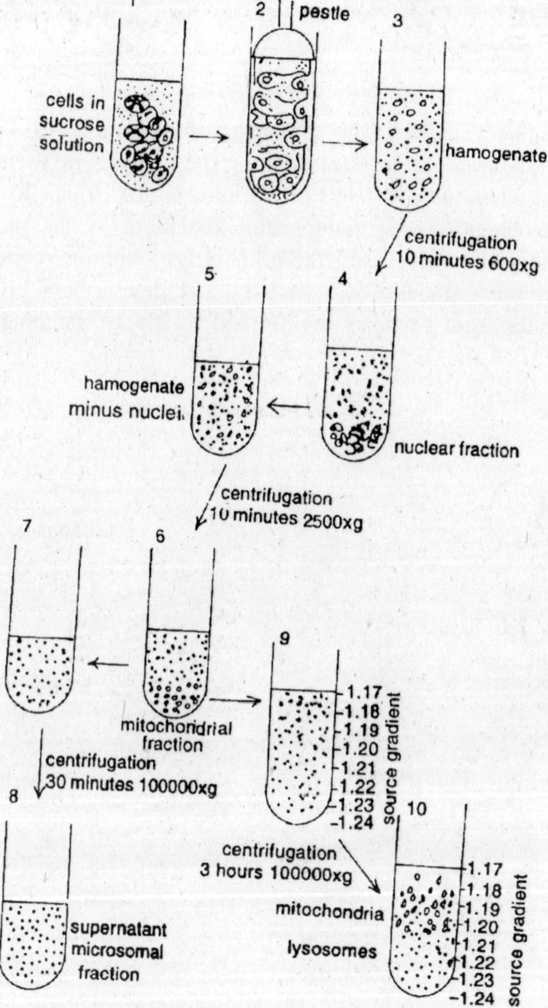

Fig. 5.4. Extraction of lysosomes by centrifugation.

the lysosomes has been developed in laboratory of *de Duve*. For the extraction of lysosome, first of all the cells are homogenated in sucrose solution. Rapid mechanical rotation of the pestle ruptures the cells, setting the intracellular particles free in the medium. Then successive centrifugation of the resulting homogenate produces fractions, which can be separated by micro-needle. The entire process of centrifugation for lysosome can be studied as shown in the Figure 5.4.

Chemistry of Lysosomes

Lysosomes contain variety of enzymes, upto the present time about 40 enzymes have been isolated in variety of tissue type. Some common enzymes, and β-galactosidase, β-glucuronidase, β-N-acetyl-glucosaminidase, α-glucosidase, α-mannosidase; cathepsin A (Acid protease), cathepsin B (Acid protease), aryl sulphatase A, aryl sulphatase B, acid ribonuclease, acid deoxyribo-nuclease, acid phosphatase, acid lipase, phospholipase A, phosphotidic acid phosphatase hyaluronidase, phosphoprotein phosphatase, amino peptidase A, dextranase, sacchariase, lysozyme (muramidase), Mg^{++} activated ATPase, indoxylacetate, esterase and plasminogen activator. All these enzymes of the lysosomes are enclosed within the single lipoprotein membrane. Most of these enzymes function more efficiently under slightly acidic medium, *pH* optima around 5.0, as such they are collectively called as acid hydrolases.

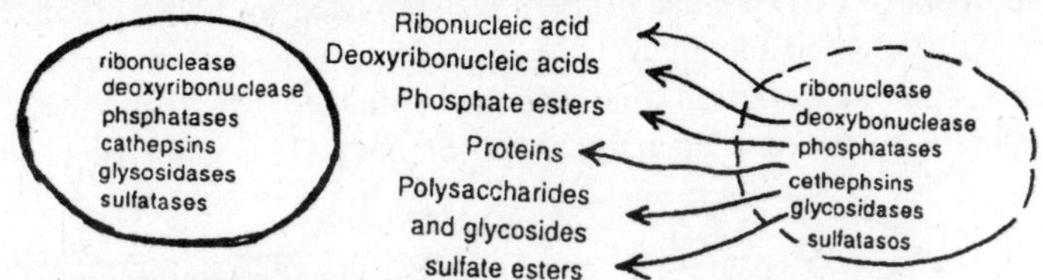

Fig. 5.5. The enzyme of lysosomes and their substances.

Different Nomenclature

The cytochemical defination of lysosomes based on the presence of a single unit membrane and a positive staining reaction for acid phosphatase and some related enzymes, may be considered for most practical purposes equivalent to the biochemical defination. As our knowledge of the singificance of lysosomes in cell physiology has progressed, it has become evident that the term lysosomes covers a variety of different forms which can be distinguished on the basis of morphological and functional criteria. The following are the terms commonly used in the literature:

(i) *Autophagic Vacuoles*. It is membrane lined vacuole containing morphologically recognizable cytoplasmic components.

(ii) *Cytolysome.* Same as Autophagic vacuoles.

(iii) *Cytosome.* Particles referred to cytosomes are usually lysosomes. Some workers include the un-related micro-bodies under this term.

(iv) *Cytoergosome.* Same as Autophagic vacuoles.

(v) *Microbody.* A particle found in liver and kidney, bounded by a single unit membrane and containing a finely granulated material. According to *de Duve,* they are definitely not lysosomes.

(vi) *Multivesicular Bodies.* Structures lined by a single membrane and containing inner vesicles resembling Golgi complex and are considered to be lysosomes.

(vii) *Residual bodies.* Membrane lined inclusions characterized by undigested residues comprises telo-lysosomes and hypothetical post-lysosomes.

FUNCTIONS

Lysosomal Digestion of External Particles

Large molecules are taken into the cell by the process called *phagocytosis*. This is a perfectly adequate and accurate term that implies a cell that eats. But recently the new term *endocytosis* has won favour. The first clear indication of a relationship between lysosomes and engulfment of extracellular material was provided by *Stians* (1952, 54, 55). As shown in the figure, the cell engulfs the particles and then forms and invagination that becomes pinched off from the cell membrane to become an internal sac or body. It is referred to as a

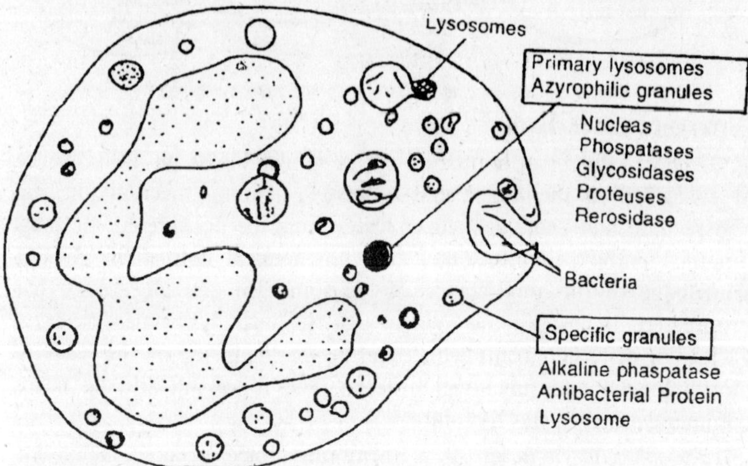

Fig. 5.6. A professional phagocyte.

phagosome. A phagosome then moves towards the lysosome. Exposure of the material to the lysosomal hydrolases occurs through fusion of the phagosome with a lysosome. This results in the formation of a *secondary lysosome* or *digestive vacuole*. The lysosome involved in the process may be primary or secondary, depending on the relative sizes of the two partners. The process may appear to an observer as a lysosome discharging enzymes into a phagosome or as a phagosome shedding its contents into a lysosome, as may be the case in hepatic parenchymal cells; or simply as a mutual sharing of the contents of the two vacuoles, if they are of a comparable sizes. Now the enzymes from the lysosome can come into contact with the molecules brought into the cell in the phagosome and digestion occurs. Once the molecules are digested, the digested products can diffuse out of the so-called digestive vacuole into the cytoplasm of the cell leaving the residue in the digestive vacuole. The digestive vacuole now moves on to the cell membrane where the so-called reverse phagocytosis or defecation occurs.

Digestion of Intracellular Substance

In certain cases portions of the cell somehow, find their way inside the cell's own lysosomes and are broken down. This process is termed as *cellular autophagy*. How do they get in, is not clear, and the role which autophagy plays in cell function can only be summarized. Proteins, fats and polysaccharides can all be synthesized, and stored in the cell. During the starvation of cell these stored food materials are digested by lysosomes to give energy. What stimulates autophagy to take place and how do the large molecules get into the lysosome is not clear.

Cellular Digestion

When a cell dies, the lysosomal membrane ruptures. The liberated enzymes become free in the cell, which then quickly digested the entire cell. The hypothesis has been advanced that this is a built in mechanism for removing dead cells. In multicellular animals, many cells are constantly being formed, live for a short period of time, and then die. The self-digestion may occur as a patho-logical mechanism just for example, it a cell is cut of from its oxygen supply or poisoned, the lysosomal membrane may rupture, thereby permitting the enzymes to dissolve the cell. Therefore, they are also con-sidered as *suicide bags* of the cells.

Extracellular Digestion

A cell can discharge lysosomal enzymes to destroy the surrounding structures. This function is performed by reverse phagocytosis. A pocket

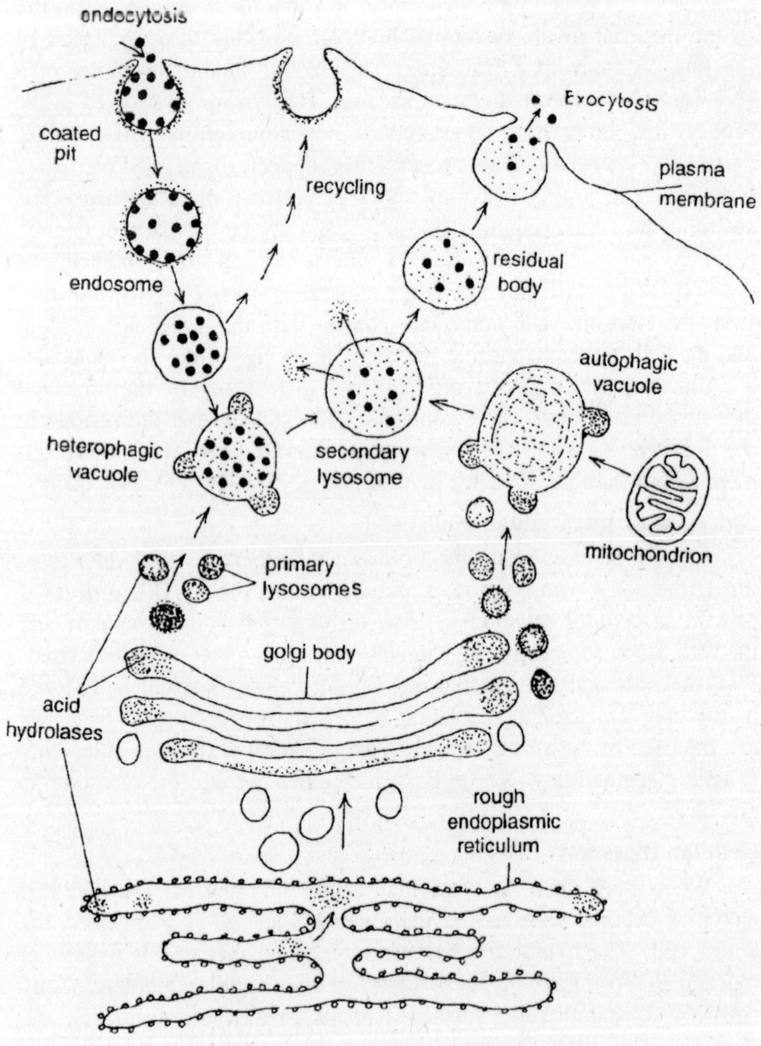

Fig. 5.7. Formation and functions of lysosomes in cellular heterophagy and autophagy.

of enz-ymes from a lysosome is released outside the cell where it then digests contigious structures. This is thought to explain, how sperms penetrate the protective coating of the ovum during fertilization. It may also explain how oesteoblast cells that destroy bones, function. This may also be explanation for the well-known ability of white blood cells to pass quickly out of the blood vessels and into thetissues spaces at the site of an infection.

Role in Secretion

In recent years evidences have started acumulating suggesting the role of lysosomes in the formation of secretory products in secretory cells. The phenomenon of lysosomes-mediated thyroid hormones secretion is the best known example of direct lysosomes involvement in secretory process. Lysosomes also play possible role in the regulation of hormone secretion. It is believed that mammotrophic hormones of the anterior pituitary is synthesized on the ribosomes of the RER and is packaged into secretory granules by passage through the Golgi.

The eptihelial cells of the thyroid also contain lysosomes rich in lysosomal enzymes. The follicles of the thyroid gland contain high molecular weight protein thyroglobulin, which is stored as colloid in the lumen. The thyroid hormones thyroxine and triiodothyroxin are linked with this protein. The colloid containing thyroglobulin enters the epithelial cell by pinocytosis. The colloid droplets fuse with primary lysosomes to form secondary lysosomes or digestive vacuoles. The thyroid hormones are split from the thyroglobulin and released into the blood stream. Thus, the thyroid hormones are released by hydrolysis of thyroglobulin.

Chromosome Breaks

Lysosomes contain the enzyme deoxyribonuclease (DNAse). This enzyme causes chromosomal breaks and their rearrangement. DNAse has two active sites and breaks down both the strands of DNA. The breaks have been produced experimentally in isolated chromosomes incubated in DNAse. These breaks lead to various syndromes.

Role in Development and Metamorphosis

Lysosomes are important in development. Good evidences have accumulated on the role of lysosomes in involution of uterus and mammary glands immediately in postpartum. During metamorphosis, the process of resorption of the tadpole tail and regression of the various larval tissues, including the fat body and the salivary gland, are accompanied by increased acid hydrolase activity.

Osteogenesis

During conversion of cartilage into bone, the special osteoblast cells produce lytic substances which erode the matrix of the cartilage and help in the formation of bone.

Role of Lysosomes during Cell Division

During the cell division, the lysosomes of that particular dividing cell move towards the periphery instead of near the nucleus, as in usual cases they are seen. During the cytokinesis roughly equal number of them move towards opposite poles. Sometimes during cell-division certain repressors in cytoplasm inhibit cell-division. Lysosomes secrete certain depressors which destroy the repressor and results in cell division.

Help in Protein Synthesis

Novikoff and *Essner* (1960) have suggested the possible role of lysosomes in protein synthesis. Recently, *Singh* (1972) has correlated lysosomes activity with the protein synthesis. In the liver and pancreas of some birds, lysosomes seem to be more active and developed showing possible relationship with cell metabolism.

Lysosomes and Cancer

Malignant cells are found to contain abnormal chromosomes. It is presumed that the chromosomal abnormality is caused by chromosomal breakage presumably produced by the lysosomal enzymes. The partial deletion of chromosome 21 in man is associated with the chronic myeloid leukemia (blood cancer).

Removal of Dead Cells

Hirsch and *Cohn* (1954) suggested that lysosomes help in the removal of dead cells in tissue. The lysosomal membrane ruptures in these cells, releasing the enzyme into body of cell, so that whole cell may be digested. This process of tissue degeneration (necrosis) is due to this lysosomal activity.

Fertilization

During fertilization the sperm releases hydrolytic enzymes from the acrosome vesicles. These enzymes help in the penetration of the sperm through the envelopes of the egg. Fluorescence microscopy studies of acridine-stained spermatozoa of the guinea pig show that the acrosome vesicles contain several enzymes, including hyaluronidase and proteases, which are also found in lysosomes. In fact, the acrosome vesicle has been looked upon as a *giant lysosome*. The acrosome vesicle enzymes also apparently activate the egg by breaking down its cortical granules.

Lysosomes and Disease

Inhalation of foreign particles like silica, asbestos etc., leads to inflammation and deposition of fibrous tissue in the lungs. The particles of silica or asbestos increase the permeability of lysosomal membranes and rupture of lysosomes. This leads to the lysis of lung cells resulting in their inflammation.

A metabolic disorder, the *gout*, is caused by the accumulation of sodium urate crystals in the joints. These are picked up by the phagocytes resulting in their lysosome rupture. This leads to acute inflammation and increased collagen synthesis. Pome's disease is another example which is caused due to absence of an lysosomal enzyme which hydrolize the glycogen. Thus, liver cells are engorged with glycogen.

Table 5.2. Some of the Disease Related to Genetic Abnromalities of Lysosomes are listed below

Disease	*Substance accumulated*	*Enzyme defect*
Ceramide lactoside	Ceramid lactoside	β-glactosidase
Gancher's disease	Glucocerebroside	β-glactosidase
Generalized gandliosidosis	Ganglioside GM_1	β-glactosidase
Krabbe's disease	Galactocerebroside	β-glactosidase
Meta chromatic leucodystrophy	Ceramide glactose-3 sulphate	Sulfatidase
Niemann-Pick disease	Sphingomyelin	Sphingomyelinase
Tay-Sach's disease	Ganglioside GM_2	Hexosaminidose-A
Type II glycogenosis	Glycogen	α-Glycosidase.
Fabry's disease	Ceramide trihexodise	α-Glactosidase

Origin of Lysosomes

They have multiple origin depending upon the tissue in which they are located or on their function in a specific cell.

(i) *Extracellular origin.* Lysosome may be the vacuole absorbed into cell by the process, pinocytosis. The pinocytic vacuole may later become cytoplasmic particle and thereon enzymatic activity becomes developed.

(ii) *Origin from Golgi complex.* There are evidences that lysosomes originate from the Golgi complex and represent zymogen granules. Their similar function and structure with Golgi complex support this view. Recent studies have shown that accumulation of secretory products within Golgi vacuoles leads to the formation of lysosomes, and membranes surrounding the products are derived from Golgi membrane.

(iii) Origin from ER. *Novikoff* (1965) reported that the lysosomes originate directly from granular endoplasmic reticulum, by a process of blebbing.

Microbodies (Peroxisomes and Glyoxisomes)

Under electron microscopic examination, in 1954 in the cells of kidney tubules, a membrane-bound with a dense granular structure appeared. It measured about 0.5-1.0 mm in diameter. It was named as *microbody*. In 1956, an another similar structure was described in the liver tissue. In these cells some type of crystallines core was seen within the vesicles. When cells are homogenized and subjected to differential centrifugation, microbodies sediment with small particle fraction containing the lysosomes and mitochondria. These microbodies contain different enzymes and they are involved in particular types of oxidative reactions. It has been shown that the absence of microbodies in liver and kidney tissue is associated with a fatal disease of infancy, *cerebrohepatorenal syndrome*.

Peroxisomes

The term peroxisomes is coined by *de Duve* (1969) for those microbodies which are rich in enzymes *peroxidases* and *catalases*, and produce hydrogen peroxide (H_2O_2) during their degenerative activity.

Structure

There are oval or spherical structures limited by a single membrane. They contain fine granular substances that may condense to form an opaque core or *nucleoid*. If the nucleoids are absent such peroxisomes are called *microperoxisomes*. Following enzymes have been isolated from peroxisomes. These are uric acid oxidase, D-amino acid oxidase, a-hydrolytic acid oxidase, NADH-glyoxylate reductase, NADP—isocitrate dehydrogenase catalase and glycolate oxides.

Function

In green plants peroxisomes are involved in *photorespiration*. The photorespiration occurs in two steps. In the first step-oxidation of one of a variety of substrates using molecular. Oxygen as the oxidizing agent, resulting in the formation of hydrogen peroxide (H_2O_2). These reactions are catalysed by various oxidases. Since H_2O_2 is a very reactive and toxic oxidizing agent, some means must be available in the organelles to destroy this compound so that it cannot accumulate. In the second step H_2O_2 is degraded into water and oxygen. This reaction is catalyzed by the enzyme catalase.

O_2

RH_2 | oxidases

R_2 H_2O_2

$R'H_2$ Catalase

R' O_2

$2H_2O$

(here RH_2 may be amino acid or uric acid).

Depending upon the plant and its environmental conditions, such as light, oxygen availability etc., chloroplasts have the potential to produce a 2-carbon compound celled *glycolate* the glycolate is formed as:

$$\text{Ribulose - 1, 5-diphosphate} \xrightarrow{O_2} \text{phosphoglycolate + 3, phosphoglycerate.}$$

$$\text{Phosphogycolate} \longrightarrow \text{glycolate.}$$

As soon as glycolate is formed in chloroplast, it diffuses out into peroxisomes

$$\text{Glycolate} + O_2 \xrightarrow{\text{Glycolate oxidase}} \text{Glyoxylate} + H_2O_2$$

The glyoxlate of peroxisome can be metabolized as

$$\text{Glyoxylate + Glutamic acid} \longrightarrow \text{Glycine} + \alpha\text{-ketogluteric acid.}$$

Glycine is transferred to the mitochondrion, where it is converted into serine.

$$\text{Glycine} \longrightarrow \text{Serine} + CO_2 + NH_3.$$

Origin of peroxisomes

The close morphological association between peroxisome and endoplasmic reticulum suggests that they are originated from endoplasmic reticulum and their enzymes are synthesized by the ribosomes.

Glyoxisomes

Along with peroxisomes, plants contain another organelles which, in addition to glycolate oxidase and catalase, also contains a number of other enzymes not found in animal cells. These are called *Glyoxisomes* and found only in the cells of endosperm or plant seedlings.

Glyoxysomes were described by *Briedenbach* (1967). Their structure is similar to that of peroxisomes except in place of granules, crystals are bounded by a single membrane. Two enzymes *isocitrase* and *malate synthetase* are present besides several essential enzymes of Krebs' cycle.

Function

One of the primary activities of glyoxisomes in the germinating seedlings is the conversion of stored fatty acids into carbohydrate i.e., gluconeogenesis. Since the product of fatty acid hydrolysis is acetyl Co-A, one might expect that glucose formation could occur by shutting the acetyl Co-A through the Kreb's cycle.

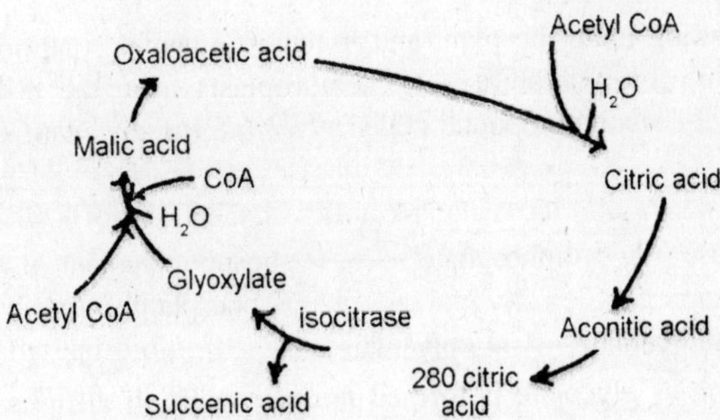

An alternate pathway for acetyl Co-A in plants cells is afforded by the *glyoxylate cycle*, which takes place in glycoxisomes. In this cycle, the first step is the condensation of acetyl Co-A with oxalo-acetic acid (as seen in Kreb's cycle), results in the formation of citric acid. Two decarboxylation steps of Kreb's cycle are by passed and succinic acid is formed along with glyoxylate. The succinic acid can be utilized in gluconeogenic reactions. Glyoxylate can serve to regenerate oxalo-acetic acid for an additional round of the cycle. Presence of catalise and glycolate oxidase provide the means for other compounds to be converted to glyoxylate, which can then enter the cycle.

6

Endoplasmic Reticulum

Endoplasmic reticulum was first of all observed in 1945 by *Porter, Claude* and *Fullam.* They noted the presence of a network or recticulum of strands associated with vesicle-like bodies in the cytoplasm of the cultural fibroblast or thinly-shaped tissue culture cells. Further electron microscopy by *Porter* and *Thompson* (1947) has revealed that these strands of reticulum are vesicular bodies inter-connected, so as to form a complex network in the inner endoplasmic part of the cytoplasm. As this network is concentrated in the endoplasm of the cell more than in the ectoplasm, therefore, it is known as *endoplasmic reticulum* (ER), or *ergastoplasm* or *vaculolar system* of the cell. The endoplasmic reticulum is not visible in the cytoplasm of a living cell under the phase-contrast microscope but the observations by electron microscope further confirmed the presence of endoplasmic recticulum as reported by *Porter* and his colleagues. Recent studies have further confirmed and accepted the concept of a structural organization of cytoplasm. Recently, under the phase-contrast microscope, *Fawcett* and *Ito* (1958), and *Rose* and *Pomerat* (1960) have studied the structure and distribution of endoplasmic reticulum in the living tissue-culture cells.

Occurrence

The endoplasmic reticulum occurs in all the eukaryotic cells except erythrocytes (R.B.Cs) of mammals. It is absent in prokaryotes. Its development varies considerably in various cell types. It is small and undifferentiated in eggs and in undiffer-entiated embryonic cells. Only a few vacuoles are present in the spermatocytes and muscle cells. However, it is highly organized in cells synthesizing proteins or in cells that are engaged in lipid metabolism.

Morphology of Endoplasmic Reticulum

The endoplasmic recticulum has been found in all kinds of mature cells except the mature mammalian erythrocyte, which is also devoid of a nucleus. Actually the first description of these structures seemed with electron microscope by *Porter, Claude* and *Fullam* in 1945 in cultured cells. These are membrane bounded sacs in the form of double membranes (cisternae) by *Sjostrand* (1953) or the name *cisternae* was given by *Sjostrand* and the name *tubules* was given by *Kurosumi* (1954). Rounded and irregular *sacs* or *vesicles* were observed by *Weiss* 1953.

Morphologically the endoplasmic reticulum is composed of following three kinds of structures, viz., 1. cisternae, 2. vesicles and 3. tubules.

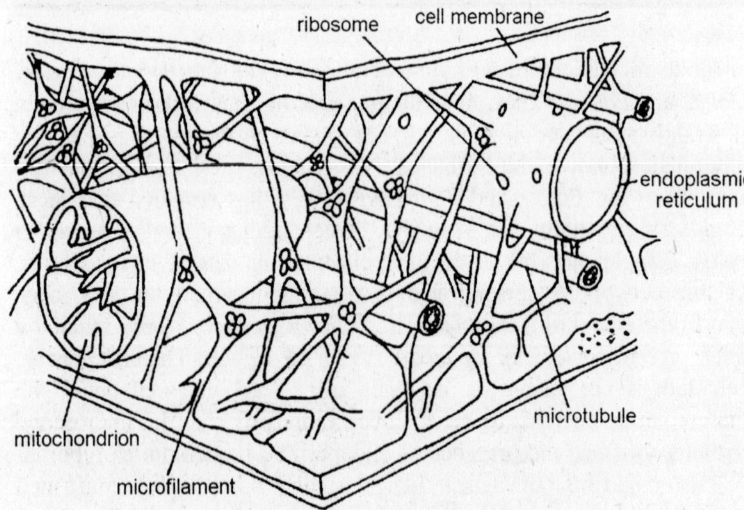

Fig. 6.1. The complex cytoskeleton of an eukaryotic cell.

Cisternae or Lamellae

They are long, flattened and usually unbranched tubules, which are arranged in parallel arrays. They are of uniform width throughout and their thickness varies from 40-50 mm. This pattern of reticulum is characteristic of basophilic regins of the cytoplasm and of those cells, which are active in protein synthesis. Lamellae or cisternae occur in the liver-cells, plasma-cells, brain-cells and in notochord cells etc.

Tubules

The tubules are small, smooth-walled, branched tubular spaces having a diameter of about 50-190 mμ. These occur in cells that are

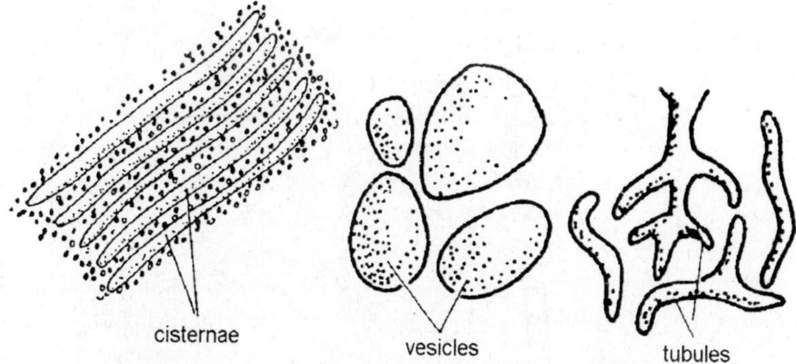

Fig. 6.2. Various components of the endoplasmic reticulum.

busy in the synthesis of steroids like cholesterol, glycerides and hormones. These are haphazardly arranged in the cytoplasm of developing spermatids of guinea pig, muscles cells and other non-secretory cells.

Vesicles

The vesicles range in diameter from 25 to 500 mμ and are for the most part rounded in shape. These are abundant in the cells engaged in the protein synthesis as in hepatic and pancreatic cells.

All these three patterns of endoplasmic reticulum may occur in the same cell or in different cells. Their arrangement also differs in different cells viz., as parallel rows in the liver-cells of mammals; haphazardly in panceratic cells or in the form of a net-work of tubules in striated muscle-cells. In notochordal cells of *Ambystoma* larva, the pattern of cisternae is of still another type.

Ultrastructure of Endoplasmic Reticulum

All the three structures of the endoplasmic reticulum are bounded by a thin membrane of 50 to 60Å thick. Like the plasma membrane, nucleus, etc., its membrane is also formed of three layers—the outer and inner dense layers are composed of protien molecules, and the two middle thin and transparent layers are of phospholipids. The endoplasmic reticular membrane is continuous with the plasma membrane, nuclear membrane and membrane of Golgi complex. The lumen of the endoplasmic reticulum acts as a passage for the secretory products and *Palade* (1956) has observed the secretory granules in it.

Types of Endoplasmic Reticulum

On the basis of presence or absence of ribosomes they are of two types:

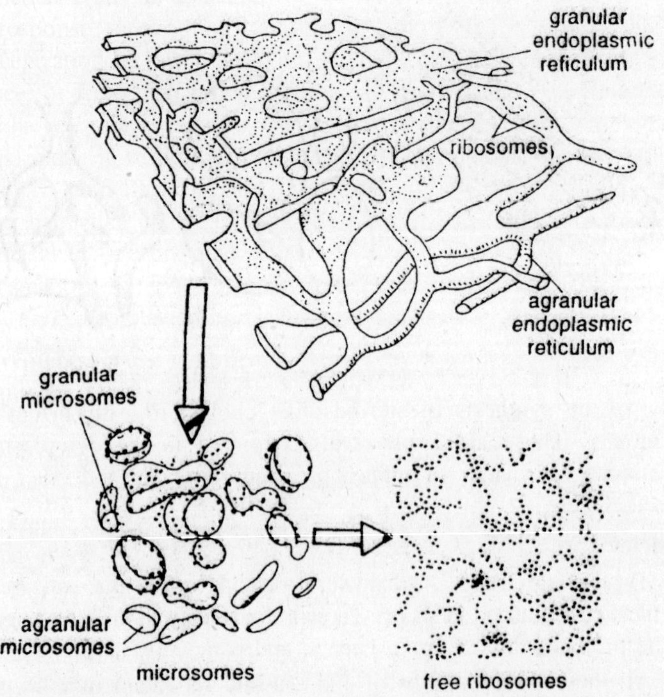

Fig. 6.3. Three-dimensional structure of endoplasmic reticulum showing microsomes and ribosomes.

Granular or Rough Walled Endoplasmic Reticulum

When the particles or *ribosomes* are present on the wall of E.R., it is called rough walled E.R. These particles are always present at the outer surface of the E.R. i.e., on the surface of the limiting membrane facing the continuous phase the matrix of the cytoplasm. The elements (E.R.) with rough surfaces are high in ribonucleic acid and are intensely basophilic. The membranes themselves are not rough, but associated with their outer surfaces are tiny particulate components 100 to 150Å in diameter. These are called *ribonucleoprotein* (RNP) *particles* or *ribosomes* and contain as the average 40% RNA and 60% protein. The elements having ribosomes are usually of the cisternal type and are found in cells active in protein synthesis. Bichemical studies have indicated that the ribosomes are important in protein synthesis, even though the membranes are not always necessary for this activity. Other functions will be explained later.

The smooth surfaced E.R. is often continuous with the rough surfaced E.R. thereby making the absence or presence of ribosomes the only significant difference between the two. The continuity between smooth and rough E.R. has been repeated by demonstration. It has been suggested more over that one grows from the other but which from which is uncertain. The ribosomes can readily be dissociated from the endoplasmic reticulum membranes by treatment with deoxycholate.

Smooth Walled Endoplasmic Reticulum

The name smooth walled is given to that portion of endoplasmic reticulum that is devoid of ribosomes. Like the rough walled endoplasmic reticulum smooth form shows a characteristic morphology which is tubular rather than cisternae. The smooth walled endoplasmic reticulum is found in the cells that are active in the synthesis of

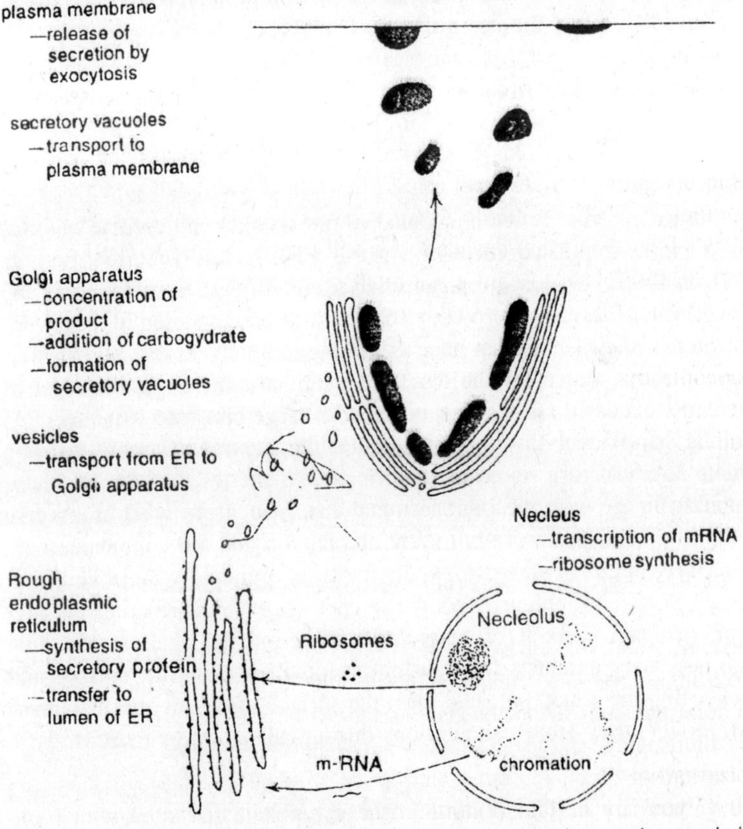

Fig. 6.4. A diagram summarizing the steps in secretion of a protein or a glycoprotein in an exocrine gland cell.

steroid compounds such as cholesterol, glycerides, and the hormones (testosterone and porgesterone). It is studied by *Fawcett* (1960) that they are also present in the pigmented epithelial cells of the retina which are involved in the metabolism of vitamin A in the production of visual pigment. Glycogen storing cells of the liver contain the smooth, tubular elements of the endoplasmic reticulum.

Table 6.1. Differences Between Rough and Smooth ER

Smooth ER	Rough ER
1. Well developed in steroid hormone secreting cells.	Well developed in protein secreting cells.
2. It tends to be tubular.	It tends to be cisternal.
3. It is less stable and autolysis readily, after the death of the cell.	It is comparatively more stable. It may persist for sometimes.
4. It is devoid of ribosomes.	Ribosomes are found associated.

Modifications of Endoplasmic Reticulum

Sarcoplasmic Reticulum

Sarcoplasmic reticulum, found in the skeletal and cardiac muscles is a highly modified form of smooth ER. It was first reported by *Veratti* (1902) as delicate plexuses in skeletal muscles surrounding the myofibrils. Electron microscopy showed it to be composed of a network of membrane like tubules which run longitudinally in the interfibrillar sarcoplasmic space for the length of each sarcomere. At the level of H and I bands, these tubules merge with large cisternal structures. At the H band level this cisterna, called the central cisterna, forms a sieve-like structure round the myofibrils. At the level of I band, these tubules merge with the large terminal cisternae, from whch transverse tubules extend peripherally to the sarcolemma and are continuous with and are deep invaginations of it. It is generally believed that sarcoplasmic reticulum plays a role not only in distributing energy-rich material needed for muscular contraction but also in providing the necessary channels for transmiting impulses along the surface and conveying the action potenital from the surface to the myofibrils within. Moreover, they store calcium ions during relaxation of muscles.

Ergastoplasm

There are certain regions in the cytoplasm that stain with basic dyes. To these regions various names have been given like chromidial substance, basoplasm, ergastoplasm and so forth. The term *ergastoplasm*

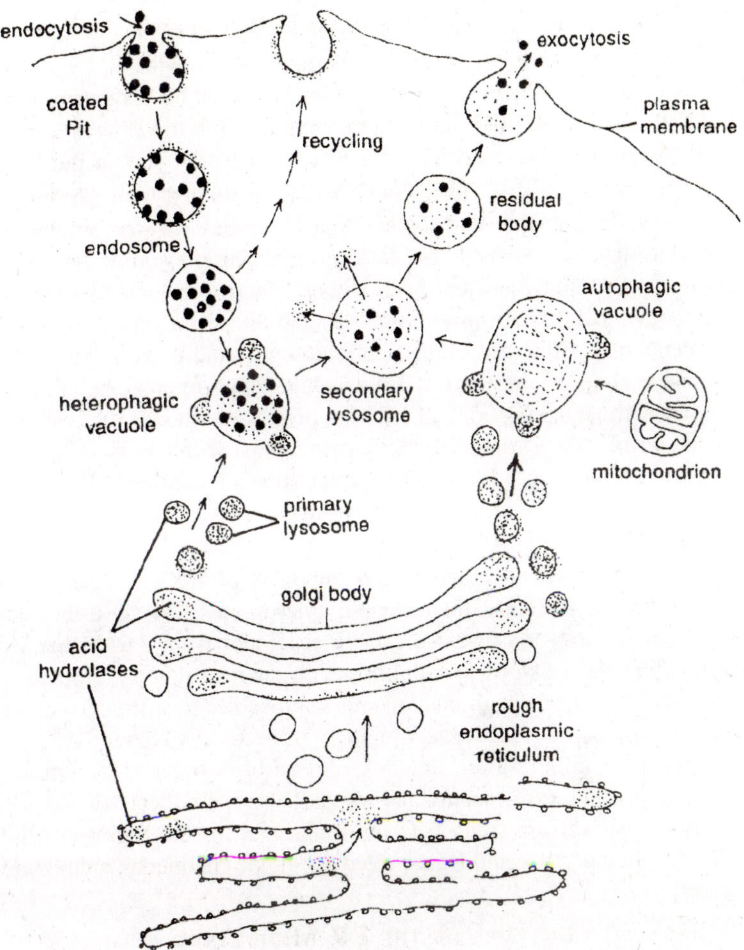

Fig. 6.5. Summary of the role of the endoplasmic reticulum and the Golgi complex in the synthesis of membrane-bond products.

was given by *Granier* in 1899 to those cytoplasmic filaments in the cells of exocrine glands which stained readily withbasic stains. *Weiss* (1953) referred to the cisternal elements as ergastoplasmic sacs. In nerve cells, such area called *Nissl bodies*. Electron microscopic studies reveal it to be an accumu!ation of ribosomes situated on the parallel lamellae of ER stalks of accumulated freely in the groundplasm. Studies of *Casperson* (1955), *Brachet* (1957) and others have demons-trated that the basophilic nature of ergastoplasm is due to the ribonucleic acid. Smooth E.R. areas of cytoplasm are never ergastoplasm.

Isolation of Endoplasmic Reticulum

Endoplasmic reticulum can also be isolated mechanically with the help of centrifuge. When tissue or cells are disrupted by homogenization the E.R. is fragmented into many smaller closed vesicles called *microsomes* (100 nm diameter), which are relatively easy to purify.

Microsomes derived from rough E.R. are studied with ribosomes and are called rough microsomes. Many vesicles of size similar to that of rough microsomes, but lacking attached ribosomes, are also found in these homogenates. Such smooth microsomes are derived in part from smooth portions of the ER and in part from vesiculated fragments of plasma membrane, Golgi complex and mitochondria (the ratio depending on the tissue). Thus, while rough microsomes can be equated with rough portions of ER, the origins of smooth microsomes cannot be so easily assigned. An outstanding exception is the liver. Because of the exceedingly large quantities of smooth ER in the hepatocyte, most of the smooth microsomes in liver homogenates are derived from smooth ER.

Ribosomes, which contain large amounts of RNA, make rough microsomes more dense than smooth microsomes. As a result, the rough and smooth microsomes can be separated from each other by sedimenting the mixture to equilibrium in sucrose density gradients. When the separated rough and smooth microsomes of a tissue such as liver are compared with respect to such properties as enzyme activity or polypeptide composition, they are remarkably similar, although not identical. It, therefore, seems that most of the components of the ER membrane can diffuse frelly between rough and smooth regions of the E.R. membrane, as would be expected for fluid, continuous membrane system.

Enzymes of the ER Membranes

The membranes of the endoplasmic reticulum are found to contain many kinds of enzymes which are needed for various important synthetic activities. The most important enzymes are the stearases, NADH-cytochrome—C—reductase, NADH dia-phorase, glucose-6-phosphotase-and-Mg^{++} activiated ATPase. Certain enzymes of the endoplasmic reticulum such as nucleotide diphosphatases are involved in the biosynthesis of phospholipid, ascorbic acid, glucuronide, steroids and hexose metabolism. The enzymes of the endoplasmic reticulum perform the following important functions:

1. Synthesis of glycerides, e.g., triglycerides, phospholipids, glycolipids.
2. Metabolism of plasmalogens.

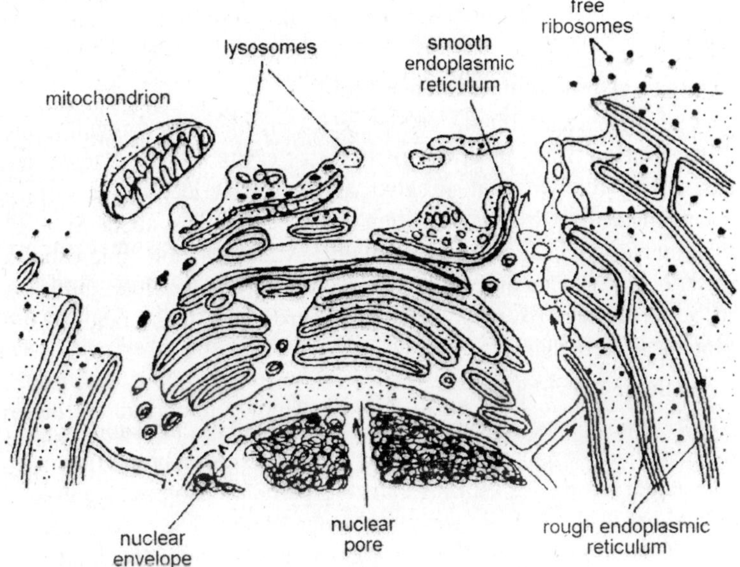

Fig. 6.6. Three-dimensional representation of cytoplasmic membrane systems and their relationship (i.e. ER, Golgi and lysosome).

3. Synthesis of fatty acids.
4. Biosynthesis of the steroids, e.g., cholesterol biosynthesis, steroid hydrogenation of unsaturated bonds.
5. $NADPH_2+O_2$—requiring steroid transformations: Aromatic hydroxylations, side-chain oxidation, deamination, thio-ether oxidations desulfuration.
6. L-ascorbic acid synthesis.
7. UDP-uronic acid metabolism.
8. UDP-glucose dephosphorylation.
9. Aryl-and steroid sulfatase.

Role of Endoplasmic Reticulum

Many functional interpretations of endoplasmic reticulum are based on the polymorphic aspects of its components in a variety of cells and its different stages of activity. More reliable interpretations are based on the same well-known facts together with hypothesis.

Mechanical Support

ER contributes to the mechanical support of the cytoplasm by dividing the fluid contents of the cell into compartments. This makes possible the existence of ionic gradient and electrical potentials along ER membranes. This concept has been specially applied to sarcoplasmic reticulum.

Exchange of Ions and any other Fluid

The membranes of the endoplasmic reticulum may regulate the exchange between the inner compartment of cavity and the cytoplasmic matrix. The following statistic gives an impressive idea of the surface area available for exchange; 1 gm of liver contains about 8 to 12 square meter of endoplasmic reticulum. After isolation, microsomes expand or shrink according to the osmotic pressure of the fluid. Diffusion and active transports may take place across the membrane of the endoplasmic reticulum.

Intracellular Circulation

The endoplasmic reticulum may act as a kind of circulatory system for intracellular circulation of various substances. Membrane flow may be an important mechanism for carrying particles, molecules and ions into and out of the cells by way of vascular system. The "pinocytosis," or "cellular drinking" also takes place by endoplasmic reticulum.

By this mechanism, particles attached to the surface of the cell or suspended in the fluid medium can be incorporated into the cytoplasm. The similar mechanism but working in a reverse direction can effect the transport of a particle from the interior of the cytoplasm to the outer medium. The continuities observed in some cases between the endoplasmic reticulum and the nuclear envelope suggest that the

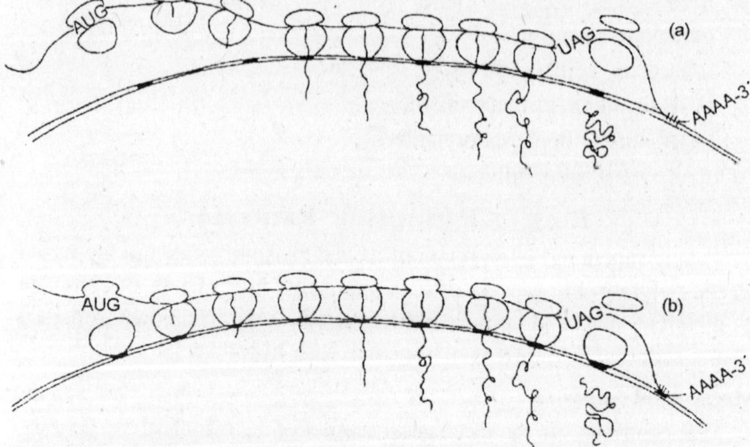

Fig. 6.7. Diagram showing the attachment of ribosomes and messenger RNA to membranes of the endoplasmic reticulum.

membrane flow may also be active at this point. This flow would provide one of the several mechanisms for export of RNA and nucleoproteins from the nucleus to the cytoplasm.

Protein Synthesis

Proteins may be synthesized to be utilized within the cell or these may have to be exported outside the cell to the site of their utility. It is the later kind of proteins in whose synthesis, endoplasmic reticulum plays an important role. For instance, rough endoplasmic reticulum, which has attached ribosomes carries synthesis of secretory proteins on these ribosomes and export them. Synthesis of tropocollagen, serum proteins and secretory granules are some examples of secretory proteins.

The protein molecules synthesized on attached ribosomes are discharged and penetrate into the cavity of ER, where they are stored or exported outside. During the transport of these products, three types of membranes ER—Golgi membrane—plasma membrane should interact and remain interconnected or disconnected due to fusion and fission respectively.

Synthesis of Lipids

The cells in which active lipid metabolism takes place are found to contain large amount of the smooth type of endoplasmic reticulum. According to some workers as *Christensen* (1961) and *Claude* (1968) the smooth type of endoplasmic reticulum is related with the synthesis and meabolism of the lipids.

Synthesis of Glycogen

The smooth endoplasmic reticulum of the glycogen storing cells of the liver and the cells of certain plants is found to be associated with the synthesis, storage and metabolism of the glycogen. But, *Porter* (1961) and *Peter* (1963) have suggested that smooth type of endoplasmic reticulum is related to glycogenolysis (breakdown of glycogen) and not to glycogenesis (synthesis of the glycogen).

Detoxification

Smooth ER is also involved in the detoxification of many endogenous and exogenous compounds. Prolonged administration of certain drugs (phenobarbitol) results in the increased activity of enzymes related to detoxification, as well as other enzymes, and a considerable hypertrophy of the SER (*Claude,* 1970). This is also applicable to administered steroid hormones.

Synthesis of Cholesterol and Steroid Hormones

Cholesterol is an important precursor of steroid hormones. The major site of cholesterol synthesis is the ER. In liver cells the SER is believed to be concerned with both the synthesis and storage of cholesterol. In the testis, ovary and the adrenal cortex the SER has a role in the synthesis of steroid hormones. The enzymes catalysing biosynthesis of androgens have been located in the SER. There is a strong correlation between the amounts of SER in cells and the capacity to synthesize steroid hormones.

Amphibian Development

There are evidences to suggest that the ER contributes in several ways to the development of the amphibian embryo.

Cell Differentiation

Some specific instance of development have been studied in detail which more or less confirm the contention that the ER is important in the process of cell differentiation. Not only this much, ER also plays role in coordinating the differentiation.

Formation of Microbodies

Closely related with the ER are microbodies, which are small granular bodies filled with the electron dense substance and limited by a single membrane. Microbodies are formed as dilations of the ER and frequently show connections with the ER cisternae. They are rich in the enzymes peroxidase (and are hence also called *peroxisomes*), *catalase*, and D-amino acid oxidase. In plant cells the enzymatic content is different and the bodies are called *glyoxysomes* because they, include enzymes of the glyoxylate cycle.

Enzyme Activities and Cellular Metabolism

Numerous enzymes mainly those involved in the metabolism of steroids (cholesterol and glycerides), phospholipids and hromones (testosterone and progesterone) are associated with the membranes of smooth endoplasmic reticulum. These membranes provide an increased inner surface for various metabolic reactions and they themselves take an active part in them by means of attached enzymes. This facilitates free union of enzymes with their substrates.

Role of Endoplasmic Reticulum in Intracellular Impulse Conduction

The existence of endoplasmic reticulum separating the cytoplasm into two compartments makes possible the existence of ionic gradients and electrical potentials across these intra cellular membranes. This idea has been applied to the sarcoplasmic reticulum, a specialized form of smooth surfaced endoplasmic reticulum found in striated muscles which is now being considered as intra-cellular conducting system. On the basis of some evidnece, it has been postulated that the sarcoplasmic reticulum transmits impulses from the surface membrane into deep regions of the muscle fibres.

Formation of Plasmodesmata

Electron microscopic studies suggest that the endoplasmic reticulum in plants plays a special role in the interconnection of cells through the cytoplasmic strands called *plasmodesmata*.

Role of E.R. During Cell Division

During cell division, some of the elements of reticulum contribute in the formation of the new nuclear membrane after karyogamy. The nuclear membrane break up into fragments in the early part of the division which finally disintigrate into small vesicles. These vesicles move towards the pole of the spindle as the metaphase starts, where they are indistinguish-able from the elements of ER. From the polar ends of the cell, elements of ER as well as the fragmented vesicles migrate into the regions around the chromosomes, which are grouping at the poles. Most of these elements of ER join or fuse around each group of daughter chromosomes to form a new nuclear envelope.

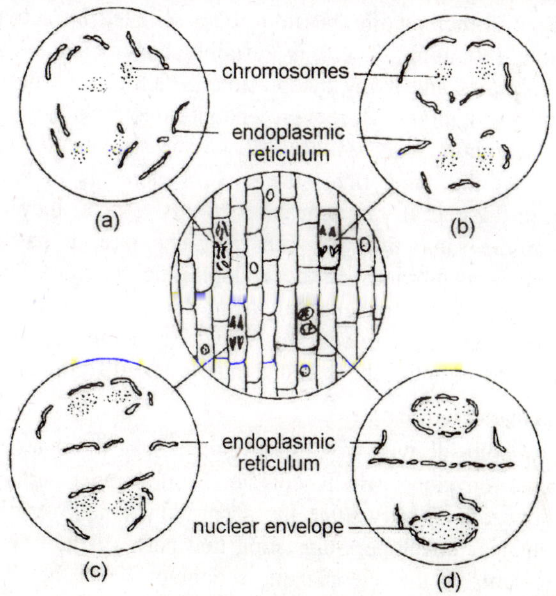

Fig. 6.8. Formation of nuclear envelope by the endoplasmic reticulum.

Transportation of Message from Genetic Material

ER provides passage for the genetic material to pass from the nucleus to the various organelles in the cytoplasm, thereby controlling the synthesis of proteins, fats and carbohydrates.

ATP Synthesis

ER membranes are the sites of ATP synthesis in the cell. The ATP is used as a source of energy for all the intracellular metabolism and transport of materials.

Formation of Cell Organelles

Most of cell organelles like Golgi complex, mitochondria, lysosomes, nuclear membrane and cell plate etc., are usually developed from endoplasmic reticulum.

ORIGIN OF ENDOPLASMIC RETICULUM

Multi step Mechanism

Although the origin of the new endoplasmic reticulum membrane has not been fully understood. The views are there. In fact, one of the possible function attributed to the endoplasmic reticulum is that of membrane biosynthesis. The protein components of endoplasmic reticulum and other membranes may be assembled by activity of the endoplasmic reticulum. There is certainly convincing evidence that Golgi membranes and many cytoplasmic vesicles an be derived from endoplasmic reticulum. Moreover, endoplasmic reticular membranes appear to be continuously synthesized, having a relatively high rate of turn over. At the same time, the several elements of endoplasmic reticulum in the cell are asynchronous in this respect, they are not all replaced at the same time or with the same rate. It has also been suggested that membranes of the endoplasmic reticulum are formed not from pre-existing elements but from the ground substance of the cytoplasm. Thus, the process by which a membrane is modified chemically and structurally is called membrane differentiation.

From Nuclear Membrane

The vacuoles derived from the evagination of the outer membrane of the nuclear envelope, which separates from its inner partner, leaving cavities between. Shortly after the separation, small vesicles appear near the nuclear envelope, suggesting that parts of the envelope give rise to elements of the endoplasmic reticulum. Thus the endoplasmic reticulum seems to have its origin in the nuclear envelope in undifferentiated cells.

7

Ribosomes

Ribosomes are the basophilic granules present in the cytoplasm of the cell. Since this material had an affinity far basic stains similar to that of chromatin granules of the nucleus, so it was for a time called *chromidial* or *chromophil substance*. The ribosomes were first noted in plant cells by *Robinson* and *Brown* in 1953 while studying been roots with the electron microscope and shortly afterward *Palade* (1955) observed them in animal cells. He isolated the ribosomes and detected the RNA in them that is why also called RNP or *ribonucleoprotein particles* or *Palade granules*. *Claude* named them as *microsome* but the name *ribosome* was designated by *Robert* in 1958.

Occurrence

They are universally distributed throughout the animal kingdom and plant kingdom. In prokaryotes too, they are found. The only cell types devoid of ribosomes is the mammalian RBC. The density of ribosomes per unit area being rather constant for any given type. It is high in the cells which are active in protein synthesis and low in cells where protein synthesis is low.

Distribution

In prokaryotic cells the ribosomes often occur freely in the cytoplasm. In eukaryotic cells the ribosomes either occur freely in the cytoplasm or remain attached to the outer surface of the membrane of endoplasmic reticulum. When they are not attached to the ER they are called free ribosomes. Free ribosomes serve as sites for the synthesis of proteins that are required to maintain the enzyme constitution of the cytoplasmic matrix.

Method of Isolation

The ribosomes are usually isolated from the cell by the differential centrifugation method in which analytical centrifuge is employed. The sedimentation coefficient of the ribosomes is determined by the various optical and electronic techniques. The sedimentation coefficient is expressed in the *Svedberg unit*, e.g., 'S' unit. The S is related with the size and molecular weight of the ribosomal particles.

Number and Concentration of Ribosomes

In all cells that contain endoplasmic reticulum, a good number of ribosomes can be observed. For example, at the base of gland cells, in plasma and liver cells, in all rapidly growing plant and animal cells and in bacteria, quantity of RNA can be related with the concentration of ribosome. In reticulocytes of rabbit nearly 100 ribosomes are found per $\mu 3$, which corresponds to 1×10^5 particles per reticulocyte and about .5% of total volume of the cell mass, or about 20,000 to 30,000 per cell. However, it the rate of protein synthesis is slowed down by unfavourable nutritional conditions, the number of ribosomes can drop considerably in protein synthesizing cells and in bacteria.

Structure of Ribosomes

Ribosomes are remarkable for their uniformity in size and composition through the wide range of cells in which they have been studied. Ribosomes of higher plants and animals are oblate, spheroids

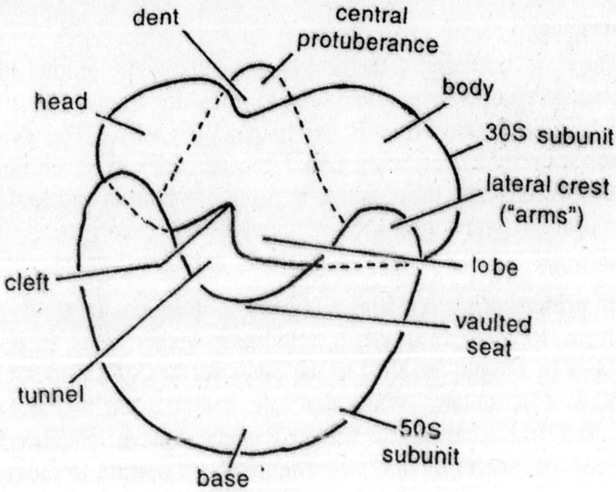

Fig. 7.1. Stoffler and Wittman's model of 70S ribosome.

and their diameter is approximately 250Å. However, the ribosomes of bacteria are some what smaller, because they contain less amount of proteins than the higher animal ribosomes. However, in quantity of RNA per particle, bacterial ribosomes resemble all the other presently studied ribosomes.

In the electron microscopic studies negative staining reveals a cleft that divides the ribosomes into a larger sub-unit and a smaller sub-unit. In *E. coli* the larger particle is somewhat 'cup' shaped or dome shaped (140 to 160Å) and smaller one forms a 'cup' (90 to 110Å) that is applied to the flat surface of the other (*Huxley* and *Zubey* 1960). In higher animals and plants it was shown that the ribosomes are attached to endoplasmic reticulum by the large sub-units.

The fine structure of the ribosome is very complex and not yet fully elucidated. Since the ribosomes are highly porous and hydrated, the RNA and protein are probably interwined within the two subunits. In sections, stained with uranyl ions (RNA selective stain), each ribosome appear as a star shaped body with four to six arms implanted on a dense axis. The isolated 50S subunit of *Bacillus subtilis* appears as a compact particle of 160 to 180Å having a pentagonal face, in the

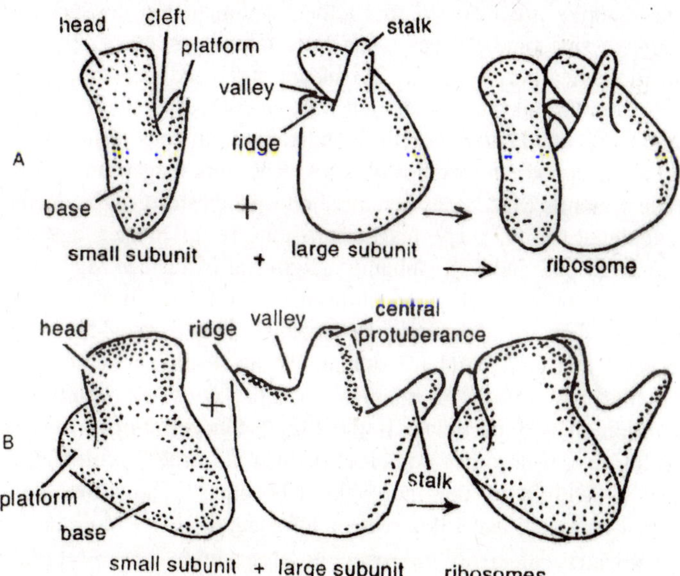

Fig. 7.2. Lake's model of the 70S ribosome. Note the three dimensional structure of the ribosome in two different orientations.

center of which is a round area of 40 to 60Å (*Nanninga,* 1967). An electron transparent core, which is negative stained ribosomes correspond to an electron opaque region, has been described in the large subunits. The 40S subunit is not regular and tends to be sub-divided into two portions that are inter-connected by a strand 30 to 60Å thick.

Ribosomal subunit

Another property which is common to all ribosomes in their subunit structure and their sedimentation constant by ultracentrifugation. On the basis of sedimentation constant there are two main types of ribosomes. Those form bacteria usually have a coefficient of 70S (S-Svedberg unit) corresponding to a molecular weight of 2.7×10^6. The others are the ribosomes of eukaryotic (nucleated cells, either plant or animal) such ribosomes posses sedimentation constant of 80S with a molecular weight of about 4×10^6 daltons.

The two, above mentioned (cup and cap) structural subunits of ribosomes require low concentration of Mg^{++} ions (0.001M) for structural cohesion. The ribosomes may be cleaned by removal of Mg^{++} ions to yield two smaller particles one 2/3 and other 1/3 of the mass of the original intact ribosome. Actually they are nothing but same subunits which are observed with the help of electron microscope. These subunits are referred to by their sedimentaion constants. The one unit of ribosome possesses a sedimentaion constant of 80S or 70S as mentioned earlier, while the 2/3 subunit has a sedimentation constant of 60S or 50S and the 1/3 subunit 40S or 30S respectively. Both the subunits are bound together through magnesium ions which interact with the phosphodiester groups of RNA. The fully magnesium saturated particle contains Mg^{++} ions per three phosphodiester groups. Removal of calculated 1/3 of these magnesium ions result in cleavage of the 80S to form 60S and 40S (subunits). Removal of further Mg^{++} results in the case of pea and yeast ribosomes at least in further cleavage liberating what are apparently 1/6 subunit. This further cleavage, in contrast to that to 2/3 and 1/3 subunits is, however, irreversible in the sense that restoration of magnesium ions, does not result in the restoration of 80S particles. If the Mg^{++} concentration is increased ten folds, two ribosomes combine to form a "dimer" with twice the molecular weight of the individual ribosomes. The dimer can be convered back into two ribosomes by lowering the Mg^{++} concentration.

The early electron microscopic observation of cells sections revealed that ribosomes were frequently associated in groups occasionally

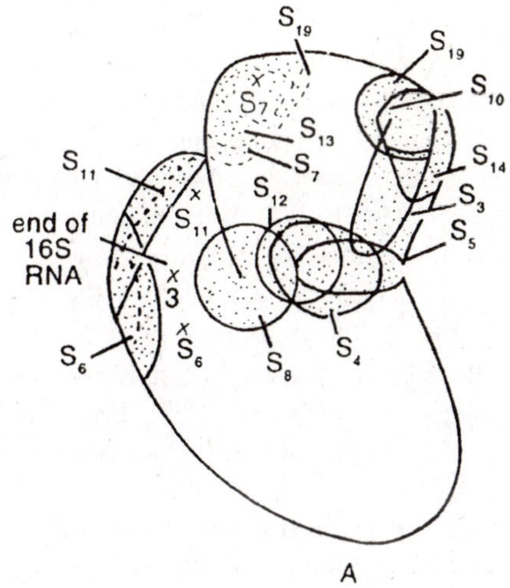

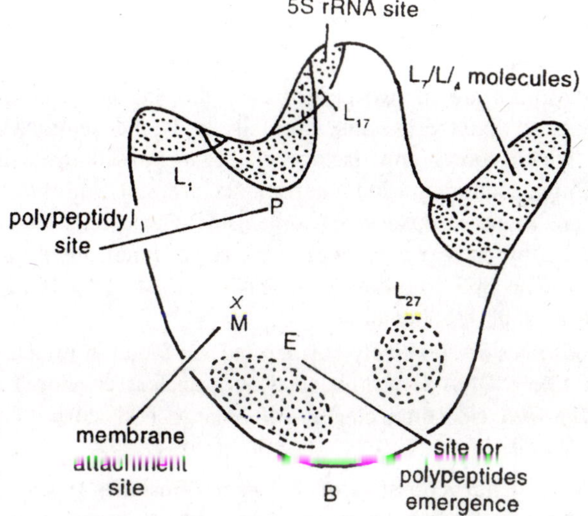

Fig. 7.3. Map of proteins in the ribosome showing their position in small subunit (A) and large subunit (B).

forming recurrent pattern. It was not until 1962 that the function of these polyribosomes, or polysomes, in protein synthesis was discovered (*Warner* and *Rich,* 1962). After treating reticulocytes with C^{14} labelled amino acids and using gentle methods for disruption, it was found that in addition to the typical sedimentation constant to single ribosome

(80S) some larger units were present. The sedimentation constant of these particles vary from 108S to 170S or even more; this corresponded to a polyribosome of five units (a pentamer). It was confirmed by electron microscopy that about 75% of the ribosomes which show 170S peak, were presented as pentameres. A thin filament, interpreted as in mRNA, about 1500Å. The number of ribosomes in a polyribosomes may vary considerably and seems to be related to the length of the mRNA that should be 'read' in the translation process.

In *E. coli* and in cells from chick embryo, polyribosomes composed of about 50 units have been observed (*Rich* 1967). Polyribosomes may be free in the cytoplasm or bound to the membranes of the endoplasmic reticulum. For free polyribosomes, a helical configuration has been postulated with the small subunits arranged around the central axis and the large subunits disposed at the periphery. In sections of various cells polyribosomes have been observed in a helical array (*Weiss* and *Grover,* 1968). It is assumed that in the polysome the mRNA is situated in between the two subunits of the ribosome.

Types of Ribosomes

The ribosomes are of two basic types, 70S and 80S ribosomes. The S refers to Svedberg units. Actually it is the sedimentation coefficient which shows how fast a cell organelle sediments in an ultracentrifuge. Sedimentation coefficients are not additive. 80S ribosomes are found in eukaryotes (organisms whose cells have true nuclei bounded by nuclear envelopes) e.g., algae, fungi, higher plants and animals. The 80S ribosome of animals consists of a large 60S subunit and a small 40S subunit.

70S ribosomes are relatively smaller and are found in prokaryotes (organisms whose DNA is not bounded by a nuclear envelope) e.g., bacteria. The 70S ribosome consists of a large 50S subunit and a small 30S subunit.

Table 7.1 Characteristics of different ribosomes types.

Ribosome Type	*Found in*	*Large subunit*		*Small subunit*	
		Size	*RNAs*	*Size*	*RNAs*
80S	Eukaryotes (animal)	60S	28-29S+5S+5.8S	40S	18S
80S	Eukaryotes (plants)	60S	25S+5S+5.8S	40S	16-18S
70S	Prokaryotes	50S	23S+5S	30S	16S
55S	Mitochondria	40S	16-17S+5S	30S	12-13S

Table 7.2. Differences between 70S and 80S ribosomes.

Presence	In Prokaryotes *(bacteria)*	In eukaroytes (algae, fungi, *higher plants and animals)*
Sedimentation Coefficient	64S-72S (average 69S)	79-85S in fungi 80S in mammals.
Size	Relatively smaller	Relatively larger
Molecular wt.	3 x 10^6	4 - 5 x 10^6
Subunits	Small 30S and large 50S	Small 40S and large 60S.
RNA	3 molecules of RNA: 16S RNA in 30S subunit, 23S and 5S RNA in 50S subunit	4 molecules of RNA 16S-18S RNA in 40S subunit; 25-29S, 5.8S and 5S RNA in 60S subunit.
M.Wt. of RNA	16S RNA-550,000 23S RNA-1,100,000 5S RNA-40,000	18S RNA-700,00 28S RNA-1,700,000 5.8S RNA-51,000 5S RNA-39,000
No. of proteins	21 (S1-S21) in small subunit 34 (S1-L34) in large subunit Total in prokaryotes: 50-60 proteins.	33 in small subunit 49 in large subunit Total in eukaryotest 70-80 proteins
Average M.Wt of protein	18,000	21,000
No of amino acids	8,000	16,000
RNA-Protein ratio	2.1	1:1

Ribosomes found in mitochondria and chloroplasts of eukaryotes are closer to prokaryote ribosomes rather than the 80S eukaryote ribosomes. Vertebrate mitochondria, for example, contain 55S ribosomes, each with a large 40S subunit and a small 30S subunit. The sedimentation coefficients 80S, 70S and 55S are rounded off values. Actual S values in different organisms may be slightly higher or lower.

Chemical Composition

The main constituents of ribosomes are RNA and proteins. The lipids are entirely absent present in traces. The ribosomes of *E. coli* passes nearly 60-65% of RNA and 35-40% protein of their weight. Table 7.3 is showing approximate quantity of RNA and protein in different types of ribosomes.

Ribosomal RNA differ in size and base contents from tRNA and from other RNA classes of most cells. Two types of RNAs are found in all ribosomes. They are an integral component, and cannot be removed easily. RNA of rat liver particles contain chiefly the common bases adenine, guanine, cytosine, and uracil with small amount of pseudouridine, the usual bases which occur in the soluble RNA are found in particles RNA only in very small amount.

Table 7.3. Showing the chemical composition of ribosomes

Ribosomes	*Sedimentation constant*	*Subunits*	*S.U. of % RNA*	*RNA in complete ribosomes*	*% protein in complete Ribosomes*
Liver	80S	60S	28S	45%	55%
Yeast	80S				
Neurospora	80S	40S	18S		
Mitochondria	70S	50S	23S		
E.coli	70S			65%	35%
Chloroplasts	70S	30S	16S		

Amino acids

The amino acid composition of trichloracetic acid insoluble protein of rat liver RNP has been determined by *Crompton* and *Petermann* (1959). Aromatic and sulphur containing amino acids were present in very small amount while leusine and arginine were high over 10%. The composition of both rabbit reticulocytes and pea seedling RNPs are similar remarkably.

A hydrochloric extract of partially purified rat live RNP has yielded arginine (*Butter,* 1960), but whether these aminoacids were derived from the proteins itself has not yet been determined.

Protein

Protein contains linear chain of amino acids. The similar in ribosome composition of amino acids of ribosomal protein is remarkably of different origin and may be quite different from that of protein formed by the ribosome. Therefore, it is common to distinguish between ribosomal structural protein and growing peptide chain of enzyme under going fabrication by the ribosome that protein is associated with ribosomal RNA by hydrogen bonding is clear from the fact that dissociation of the two is rapidly and readily accomplished by reagents which attack hydrogen bonds as for example by guanidium bromide.

Experiments have been done by *Yin* and *Bock* (1960), who obtained a stable yeast ribosomal protein. *Watson* (1960) got the protein of *E.coli* and by analysis of such a group indicates the molecular weight of each of sub-units to be about 30,000.

Ribosomal enzymatic proteins

Most of the ribosomal proteins act as enzymes thereby catalyzing protein synthesis. Initiation proteins IF1, IF2, and IF3 initiate the process of protein synthesis whereas transfer proteins (G-factor, Ts-factor) help in translocation of ribosomes over mRNA and transfer to t-RNA residue from one site of the ribosome to the other site. Another enzymes peptidyl transferase helps in the transformation of peptide chains to aminoacyl 1-tRNA and other enzymes—termination of completed polypeptide chain.

As a result of washing of ribosomes with NH_4Cl and subjected them to column chromatography, *Ochoa* and *coworkers* (1960) isolated above factors in protein synthesis. Among them, three initiation factors—IF1, IF2, and IF3—are loosely associated with 30S subunit. IF1 factor is a basic protein having molecular weight of 9200 daltons. It is involved in the binding of F-met-tRNA. IF2 factor is also a protein of mol. wt. 8000 daltons and contains-SH groups which help in binding with GTP. Third protein factor— IF3, does not require GTP and is involved in binding mRNA to 30s subunits. It is a basic proteins having molecular weight of 30,000 daltons. IF3 may also act as dissociation factor for 70s ribosomes. *Ochoa* et. al. (1972) have further reported interference factor (i) in bacteria *E. coli*. These factors bind to IF3 factor, change its specificity and thus regulate the translation of genetic message at the beginning.

Elongation factors are essential for the elongation of polypeptide chain. There are EFG (also called G factor or translocase) and EFT factor. As described earlier EFG or G factor is involved in the translocation of mRNA. In *E. coli*, it consists of single polypeptide chain having mol. wt. of 72,00 daltons. EFG + GTP promotes the translocation of newly elongated peptidyl 1-tRNA. Another EFT factors has two kinds of proteins namely Tu (temperature unstable) and Ts (temperature stable). The EFTu factor+GTP forms complex aminoacyl-tRNA before its binding to the acceptor site of ribosome being catalyzed by Ts factor. Moreover, 50s ribosomal subunit has an enzyme-peptide synthetase or peptiydl transferase associated in the formation of peptide bond. Termination factors R1 and R2 are releasing proteins helping in liberating the polypeptide chain.

Biogenesis of Ribosomes

The biogenesis of ribosomes in bacterial prokaryotic cells takes place inside the cytoplasm because of the absence of nucleous. The rRNAs originate from the specific codons of the genome or the ribosomal DNA (rDNA).

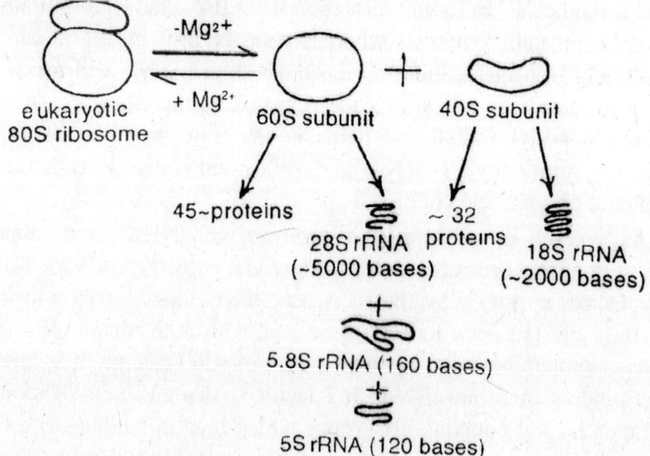

Fig. 7.4. Different RNA and protein components of eukaryotic ribosomes.

In eukaryotic cells the process of biogenesis of ribosomes is complicated and occurs inside the nucleolus. One of the chromes of a set has a specific nucleolus organizer region, which contains nucleolar RNA molecule which is a precursor of both 28S and 18S rRNA. The process of conversion of 45S-RNA into 28S and 18S rRNA is illustrated below:

45S nucleolar RNA molecules are methylated (-CH_3 group is added). These methylated molecules of 45S nucleolar RNA become associated with the necessary proteins present in the nucleolus forming 80S ribonucle-oprotein molecule particles (RNP). These 80S RNP with 45S molecular RNA split up into 32S and 18S rRNA through several intermediate steps, by which non-methylated portion of the molecules are lost. 18S molecule together with its protein molecules immediately transported to the cytoplasm. The 32S rRNA remains inside the nucleolus for sometime and then gets cleaved into 28S rRNA.

The 5S rRNA is synthesized outside the nucleolus and the genes lie adjacent to nucleolar organizer region of the chromosome.

The 18S rRNA together with its proteins leaves the nucleus through nucleopore and comes out into the cytoplasm, where in association

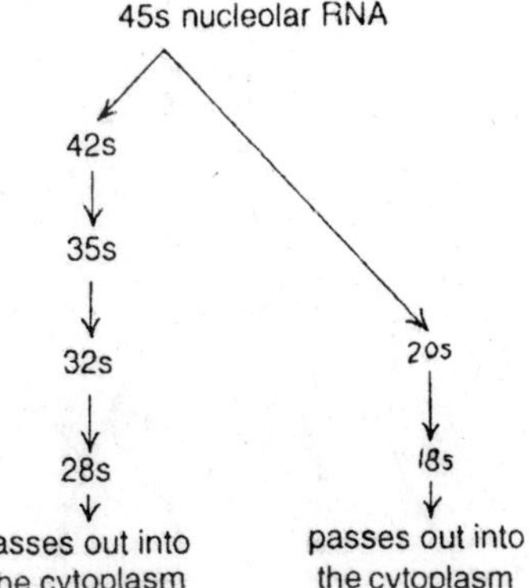

Fig. 7.5. Diagrammatic representation of the seps involved in the conversion of 40S nucleolar RNA into 28S and 18S rRNA.

with proteins this assembles into small subunit (40S) of the ribosome. The 28S rRNA also leaves the nucleus and gets incorporated with 5S rRNA and the proteins forming 60S subunit.

The synthesis of ribosomal occurs partly inside the nucleolus and partly inside the cytoplasm. The proteins synthesized in the cytoplasm assemble in the nucleolus to be utilized in ribonucleoprotein molecule particles (RNP).

Ribosomal RNA

The 70S ribosomes contain three types of rRNA, viz., 23S rRNA, 16S rRNA and 5S RNA. The 23S and 5S rRNA occur in the larger 50S ribosomal subunit, while the 16S rRNA occurs in the smaller 30S ribosomal subunit. The 23S rRNA consists of 3200 nucleotides, 16S rRNA contains 1600 nucleotides and 5S rRNA includes 120 nucleotides in it (*Brownlee,* 1968, *Fellner*, 1972).

The 80S ribosomes also contain three types of rRNA, viz., 28S rRNA, 18S rRNA and 5S rRNA. The 28S and 5S rRNA occur in the larger 60S ribosomal unit, while the 18S rRNA occur in the smaller 40S ribosomal subunit.

The 28S rRNA has the molecular weight 1.6×10^6 daltons and its molecule is double stranded and having nitrogen bases in pairs. The 18S rRNA has the molecular weight 0.6×10^6 daltons. The molecule of 5S rRNA has a clover leaf shape and a length equal to 120 nucleotides (*Forget* and *Weissmann,* 1968).

Other constituents

The metal contents or ribosomes are also a difficult question. There is little doubt that magnesium is the chief ion physiologically,

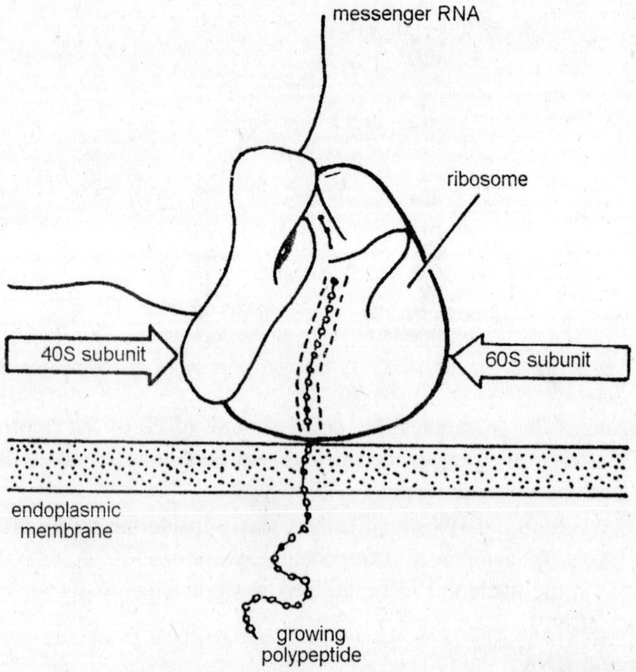

Fig. 7.6. A three dimensional model of eukaryotic cytoplasmic ribosome.

since it is present in high concentration in rat liver ribosomes. It is the most active in maintaining the 80S structure (*Hamilton* and *Petermann,* 1959) as it is essential for amino acid incorporation in *vitro* (*Rendi* and *Hultin,* 1960). The other metals which are present in the ribosomes are chromium, manganese, nickel, iron and calcium (*Walker* and *Valle,* 1958 and *Tso* 1958).

Function of Ribosomes

It is a well-known fact that the ribosomes are factories of manufacturing of protein in a cell but actually one ribosome cannot take part in the protein synthesis. It was known from 1962 when a report was published which showed that active units are no individual ribosome, but a group of these units, which is called as polyribosome. A detailed role of polyribosome was published in Scientific American Dec. 1963 by *Gric* and *Hall.* There is much evidence that ribosomes are alike and are at least interchangable.

In case of haemoglobin with a chain of 150 amino acids the most usual number of ribosomes which were linked together was 5. There may be large number of ribosomes and upto 40 ribosomes have been described. So the name polysome can also be applied at the place of polyribosome.

Protein consists of linear chain of amino acids. The chain may be short or long in both the cases it is called polypeptide chain. Polypeptide can be folded in a specific manner often combine to form a complex protein.

The ribosomes in a polysome are separated by gap of 50-150Å. Positive staing by Uracil-acetate reveals that the ribosomes are connected by a thin thread 10-15Å in diameter, which is about the thickness of a single strand of RNA. From the size of the gap between ribosomes, the total measurement of the gap containing five ribosomes, is about 1500Å, the five ribosomes have inter ribosomal gap of 50-150Å each.

8

MENDELIAN ANALYSIS

The gene is the focal point of the discipline of modern genetics. In all lines of genetic research, it is the gene that provides the common unifying thread to a great diversity of experimentation. Geneticists are concerned with the transmission of genes from generation to generation, with the nature of genes, with the variation in genes, and with the ways that genes function to dictate the features that constitute any given species.

In this chapter we trace the birth of the gene as a concept. We shall see that genetics is, in one sense, an abstract science: most of its entities began as hypothetical constructs in the minds of geneticists and were later identified in physical form if the reasoning was sound.

The concept of the gene (but not the word) was first set forth in 1865 by Gregor Mendel. Until then, little progress had been made in understanding heredity. The prevailing notion was that the spermatozoon and egg contain a sampling of essences from the various parts of the parental body; at conception, these essences are somehow blended to form the pattern for the new individual. This idea of *blending inheritance* evolved to account for the fact that offspring typically show some characteristics similar to those of both parents. However, attempts to expand and improve this theory never led to much progress.

As a result of his research with pea plants, Mendel proposed instead a theory of *particulate inheritance*. A genetic determinant of a specific character is passed on from one generation to the next as a unit, without any blending of the units. This model explained many observations that could not be explained by blending inheritance. It also proved a very fruitful framework for further progress in understanding the mechanism of heredity. For many reasons, the importance of Mendel's ideas was not recognized until about 1900 (after his death). His report was then rediscovered by three scientists after each independently had reached the same conclusions. Historically, then, Mendel's work was irrelevant to the development of ideas about heredity. However, his achievement is important, and his analysis provides a good example of basic genetic reasoning, so we now examine Mendel's work.

Mendel's Experiments

Mendel's studies provide an outstanding example of good scientific technique. He chose research material well suited to study of the problem at hand, designed his experiments carefully, collected large amounts of data, and used mathematical analysis to show that the results were consistent with his explanatory hypothesis. The predictions of the hypothesis were then tested in a new round of experimentation. (Some historians of science believe that Mendel may have "fudged" his data to fit his hypothesis, but we shall accept his reports at face value in our discussion here.)

Mendel studied the garden pea (*Pisum sativum*) for two main reasons. First, peas were available through a seed merchant in a wide array of distinct shapes and colours that are very easily identified and analyzed. Second, peas left to themselves will *self* (self-pollinate) because the male parts (anthers) and female parts (ovaries) of the flower—which produce the pollen and the eggs, respectively—are enclosed in a petal box, or keel. The gardener or experimenter can cross (cross-pollinate) any two plants at will. The anthers from one

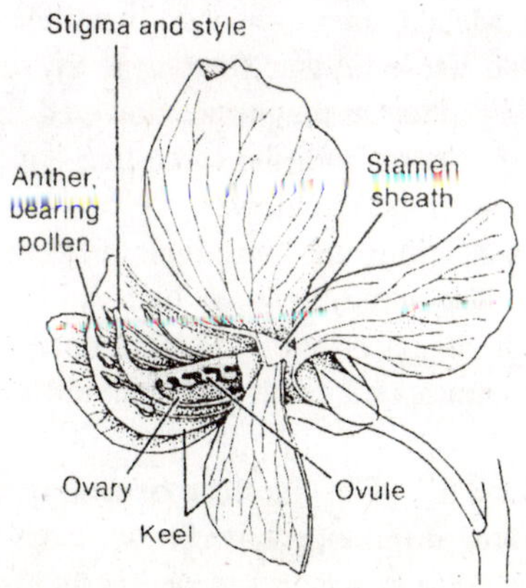

Fig. 8.1. Cutaway view of reproductive parts of a pea flower

plant are clipped off to prevent selfing; pollen from the other plant is then transferred to the receptive area with a paintbrush or on transported anthers. Thus the experimenter can readily choose to self or to cross the pea plants.

Other practical reasons for Mendel's choice of peas were that they are cheap and easy to obtain, take up little space, have a relatively short generation time, and produce many offspring. These considerations are typical of those involved in the choice of organism for any piece of genetic research. This is a crucial decision and is nearly always based upon not only scientific criteria but also a good measure of expediency.

Plants Differing in One Character

First, Mendel chose several characters to study. It is important to clarify the meaning of character in this sense. Here, a *character* is a specific property of an organism; geneticists use this term as a synonym for characteristic or trait.

For each of the characters he chose, Mendel obtained lines of plants, which he grew for two years to make sure they were pure lines. A *pure line* is a population that breeds true for the particular character being studied; that is, all offspring produced by selfing or crossing within the population show the same form for this character. By ensuring pure lines in his research material, Mendel made a clever beginning: he had established a basis of identifiable constant behaviour, so that any changes observed following deliberate manipulation in his research would be scientifically meaningful; in effect, he had set up a control experiment.

Two of the lines Mendel grew proved to breed true for the character of flower colour. One line bred true for purple flowers, and the other for white flowers. Any plant in the purple-flowered line—when selfed, or when crossed with others from the same line—produced seeds that all grew into plants with purple flowers. When these plants in turn were selfed or crossed within the line, their progeny also had purple flowers, and so on. The white-flowered line similarly produced only white flowers through all generations. Mendel obtained seven pairs of pure lines for seven characters, with each pair differing in respect to only one character.

Each pair of Mendel's plant lines can be said to show a *character difference*, a contrasting difference between two lines of organisms (or between two organisms) with respect to one particular character. The differing lines (or individuals) represent different forms that the character may take: they can be called character forms, character variants, or phenotypes. The useful term phenotype (derived from Greek) literally means "the form that is shown." The term is extensively used in genetics, and we use it in this discussion, even though such words as gene and phenotype were not coined or used by Mendel. We shall describe Mendel's results and hypotheses in terms of more modern genetic language.

Each character represented by two contrasting phenotypes. Contrasting phenotypes for a particular character are the starting point for any genetic analysis, by Mendel or by the modern geneticist. Of courses, the definition of characters is somewhat arbitrary; there are many different ways to "split up" an organism into characters. For example, consider the following different ways of staling the same character and character difference.

Character	*Phenotypes*
Flower colour	Red versus white
Flower redness	Presence versus absence
Flower whiteness	Absence versus presence

In many cases, the description chosen is a matter of convenience (or chance). Fortunately, the choice of definition does not alter the final conclusions of the analysis, except, in the names used.

We turn now to some of Mendel's specific experimental results. In our discussion, we shall follow his analysis of the lines breeding true for flower colour.

In one of his early experiments, Mendel used pollen from a white-flowered plant to pollinate a purple-flowered plant. These plants from the pure lines are called the *parental generation* (P). All the plants resulting from this cross had purple flowers. This progeny generation is called the *first filial generation* (F_1). (The subsequent generations in such an experiment are called F_2, F_3, and so on.)

Mendel also made a *reciprocal cross*. In most plants, any cross can be made in two ways, depending on which phenotype is used as male (♂) or female (♀). For example, the two crosses

phenotype A♀ × phenotype B♂

phenotype B♀ × phenotype A♂

are reciprocal crosses. Mendel's reciprocal cross, in which a white flower was pollinated by a purple-flowered plant, produced the same result in the F_1. Mendel concluded that it makes no difference which way the cross is made. If one parent is purple-flowered and the other white-flowered, all plants in the F_1 are purple-flowered. The purple flower colour in the F_1 generation is identical to that in the purple-flowered parental plants. In this case, the inheritance obviously is not a simple blending of purple and white colours to produce some intermediate colour. To maintain a theory of blending inheritance, we would have to assume that the purple colour is somehow "stronger" than the white colour, completely overwhelming any trace of the white phenotype in the blend.

Next, Mendel selfed the F_1 plants, allowing the pollen of each flower to fall on the stigma within its petal box. He obtained 929 pea seeds from this selfing (the F_2 individuals) and planted them. Interestingly, some of the resulting plants were white-flowered; the white phenotype had reappeared! Mendel then did something that, more than anything else, marks the birth of modern genetics: he *counted* the numbers of plants with each phenotype. This procedure seems obvious to modern biologists after another century of quantitative scientific research, but it had seldom if ever been used in genetic studies before Mendel's work. There were 705 purple-flowered plants and 224 white-flowered plants. Mendel observed that the ratio of 705:224 is almost a 3:1 ratio (in fact, it is 3:1:1).

Mendel repeated this breeding procedure for six other pairs of pea character differences. He found the same 3:1 ratio in the F_2 generation for each pair. By this time, he was undoubtedly beginning to believe in the significance of the 3:1 ratio and seeking an explanation. The white phenotype is completely absent in the F_1 generation, but it reappears (in its full original form) in one-tourth of the F_2 plants. It is very difficult to devise an explanation of this result in terms of blending inheritance. Even though the F_1 flowers are purple, the plants still must carry the *potential* to produce progeny with white flowers.

Table 8.1. Results of all Mendel's crosses in which parents differed for one character

Parent phenotypes	F_1	F_2	F_2 *ratio*
1. Round × wrinkled seeds	All round	5474 round; 1850 wrinkled	2.96:1
2. Yellow × green seeds	All yellow	6022 yellow; 2001 green	3.01:1
3. Purple × white petals	All purple	705 purple; 224 white	3.15:1
4. Inflated × pinched pods	All inflated	882 inflated; 299 pinched	2.95:1
5. Green × yellow pods	All green	428 green; 152 yellow	2.82:1
6. Axial × terminal flowers	All axial	651 axial; 207 terminal	3.14:1
7. Long × short stems	All long	787 long; 277 short	2.84:1

Mendel inferred that the F_1 plants receive from their parents the ability to produce both the purple phenotype and the white phenotype, and that these abilities are retained and passed on to future generations rather than blended. Why is the white phenotype not expressed in the F_1 plants? Mendel invented the terms *dominant* and *recessive* to describe this phenomenon without explaining the mechanism. In modern terms, the purple phenotype is dominant to the white phenotype, and the white phenotype is recessive to the purple. Thus, the phenotype of the F_1 provides the operational definition of dominance.

Mendel made another important observation when he individually selfed the F_2 plants. In this case, he was working with the character of seed colour. Because this character can be observed without growing plants from the peas, much larger numbers of individuals can be counted. (In this species, the colour of the seed is characteristic of the offspring—the seed itself—rather than of the parent plant.) Mendel used two pure lines of plants with yellow and green seeds, respectively. In a cross between one plant from each line, he observed that all of the F_1 peas were yellow. Symbolically,

$$
\begin{array}{ll}
P & \text{yellow} \times \text{green} \\
 & \quad\quad \downarrow \\
F_1 & \quad \text{all yellow}
\end{array}
$$

Therefore, in this character, yellow is dominant and green is recessive.

Mendel grew F_1 plants from these yellow F_1 peas and selfed the plants. Of the resulting F_2 peas, 3/4 were yellow and 1/4 were green — the 3 : 1 ratio again. He then grew plants from 519 of the yellow F_2 peas and selfed each of these F_2 plants. When the peas appeared (the F_3 generation), he found that 166 of the plants had only yellow peas. The remaining 353 plants bore both yellow and green peas on the same plant. Counting all the peas from these plants, he again obtained a 3:1 ratio of yellow to green peas. The green F_2 peas all proved to be pure-breeding green peas; that is, selfing produced only green peas in the F_3 generation.

In summary, all of the F_2 green peas were pure-breeding greens like one of the parents. Of the F_2 yellows, about 2/3 were like the F_1 yellows (producing yellow and green seeds in a 3:1 ratio when selfed), and the remaining 1/3 were like the pure-breeding yellow parent. Thus, the study of the F_3 generation revealed that the apparent 3:1 ratio in the F_2 generation could be more accurately described as a 1:2:1 ratio.

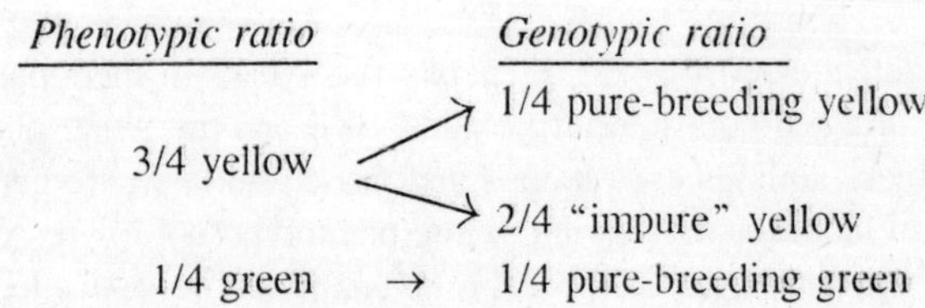

Further studies showed that these 1:2:1 ratios existed in all of the apparent 3:1 ratios that Mendel had observed. Thus the problem really was to explain the 1:2:1 ratio.

Mendel's explanation was a classic example of a model or hypothesis derived from observation, a model that was well suited for testing by further experimentation. Mendel deduced the following explanation.

1. There are hereditary determinants of a particulate nature. (He saw no blending of phenotypes, so he was forced to this particulate notion.) We now call these determinants genes.
2. Each adult pea plant has two genes—a *gene pair*—in each cell for each character studied. The reasoning here was obvious: the F_1 plants, for example, must have had one gene responsible for the dominant phenotype, called the *dominant gene*, and one gene for the recessive phenotype, which showed up only in later generations, called the *recessive gene*.
3. The members of each gene pair segregate (separate) equally into the gametes. In animals the gametes are readily identifiable as eggs and sperms. Plants produce eggs and sperms too, but these forms are less easily identified as such.
4. Consequently, each gamete carries only one member of each gene pair.
5. The union of gametes, to form the first cell (or zygote) of a new progeny individual, is random and occurs irrespective of which member of a gene pair is carried.

These points can be illustrated diagrammatically for a general case, using *A* to represent a dominant gene, and *a* the recessive gene (as Mendel did), much as a mathematician uses symbols to represent abstract entities of various kinds.

The whole model made beautiful sense of the data. However, many beautiful models have been knocked down under test: Mendel's next job was to test it. He did this by taking (for example) an F_1 yellow and crossing it with a green. A 1:1 ratio of yellow to green seeds could be predicted in the next generation. If we use *Y* to stand

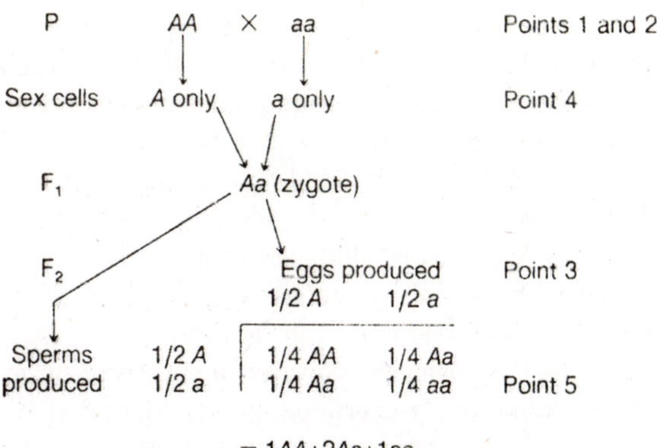

Fig. 8.2. Symbolic representation of the P, F_1, and F_2 generations in Mendel's system involving a character differene determined by one gene difference.

for the dominant gene causing yellow seeds and *y* to stand for the recessive gene causing green seeds. In this experiment, he obtained 58 yellow and 52 green seeds, a very close approximation to the predicted 1:1 ratio, and confirming the equal segregation of *Y* and *y* in the F_1 individual. This concept of *equal segregation* has been given formal recognition as Mendel's first law.

- *Mendel's First Law*: The two members of a gene pair segregate (separate) from each other into the gametes, so that one-half of the gametes carry one members of the pair and the other one-half of the gametes carry the other member of the gene pair.

Now we need to introduce some more terms. The individuals represented by *Aa* are called *heterozygotes*, or sometimes, *hybrids*, whereas those in pure lines are called *homozygotes*. Thus, an *AA* plant is said to be homozygous for the dominant gene, sometimes called *homozygous dominant*; an *aa* plant, is homozygous for the recessive

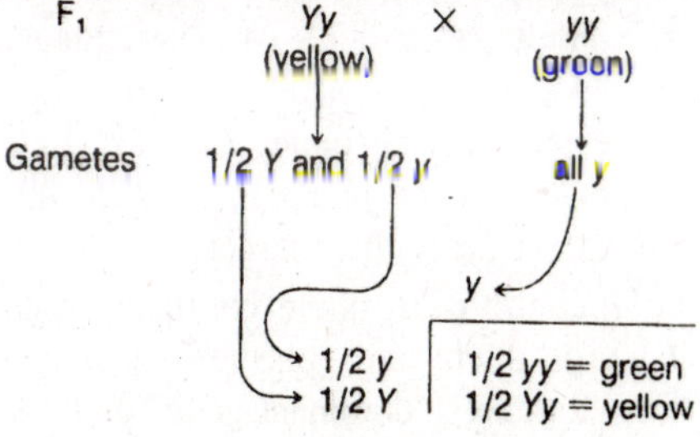

Fig. 8.3. Predicted consequences of crossing an F_1 yellow with any green.

gene, or *homozygous recessive*. The designated genetic constitution with respect to the character or characters under study is called the genotype. Thus, *YY* and *Yy,* for example, are different genotypes even though the seeds of both types are of the same phenotype (that is, yellow). You can see that in such a situation the phenotype can be thought of simply as the outward manifestation of the underlying genotype.

Note that the expressions dominant and recessive have been used in conjunction with both the phenotype *and* the gene. This is accepted usage. The dominant phenotype is established in analysis by the appearance of the F_1. Obviously, however, a phenotype (which is merely a description) cannot really exert dominance. Mendel showed that the dominance of one phenotype over another is in fact due to the dominance of one member of a gene pair over the other.

Let's pause to let the significance of this work sink in. What Mendel had done was to develop an analytical scheme for the identification of major genes regulating any biological character or function. Starting with two different phenotypes (purple and white) of one character (petal colour), he was able to show that the difference was caused by differences centering on one gene pair. Let's take the two forms of the petal-colour character as an example. Modern geneticists would say that Mendel's analysis had identified a major gene for petal colour. What does this mean? It means that in these organisms there is a *kind* of gene that has i) profound effect on the colour of the petals. This gene can exist in different forms: the dominant form of the gene (represented by *C)* causes purple petals, and the recessive form of the gene (represented by *c*) causes white petals. The forms *C* and *c* are called *alleles* (or alternative forms) of that gene for petal colour. They are given the same letter symbol to show that they are forms of the same kind of gene. One could express this another way by saying that there is a kind of gene, called phonetically a "see" gene, with alleles *C* and *c*. Any individual pea plant will always have two "see" genes, forming a gene pair, and the actual members of the gene pair can be either *CC*, *Cc,* or *cc*. Notice that although the members of a gene pair can produce different, effects, they obviously both affect the same character.

Students often find the term allele confusing, and the reason is probably that the words allele and gene are used interchangeably in some situations. For example, "dominant gene" and "dominant allele" both refer to the same thing in an interchangeable way. This stems from the fact that the forms (alleles) of any type of gene are of course genes themselves.

Plants Differing in Two Characters

The experiments described thus far dealt with a single gene pair—that is, a *monohybrid* system; we considered the allelic forms of a gene affecting one character. The next obvious question is what happens when a *dihybrid* cross is made, involving genes affecting two different characters. We can use the same symbolism that Mendel used to indicate the genotype of seed colour—*Y* and *y*—and seed shape—*R* and *r*.

A pure-breeding line of *RRyy* plants, on selling, produces seeds that are round and green. Another pure-breeding line is *rrYY*; on selfing, this line produces wrinkled yellow seeds (*r* is a recessive allele of the seed-shape gene arid produces a wrinkled seed). When Mendel crossed plants from these two lines, he obtained round yellow F_1 seeds, as expected. The results in the F_2 are complex. Mendel performed similar experiments using other pairs of characters in many other dihybrid crosses: in each case, he obtained 9:3:3:1 ratios. So, he had another phenomenon to explain, some more numbers to turn into an idea.

He first checked to see whether the ratio for each gene pair in the dihybrid cross is the same as that for a monohybrid cross. If you look at only the round and wrinkled phenotypes and add up all the seeds falling into these two classes, the totals are 315 + 108 = 423 round, and 101 + 32 = 133 wrinkled. Hence the monohybrid 3:1 ratio still prevails. Similarly, the ratio for yellow to green is (315 + 101):(108 + 32) = 416 : 140 ≅ 3: 1. From this clue, Mendel concluded that the two systems of heredity are independent. He was mathematically astute enough to realize that the 9:3:3:1 ratio is nothing more than a random combination of two independent 3:1 ratios.

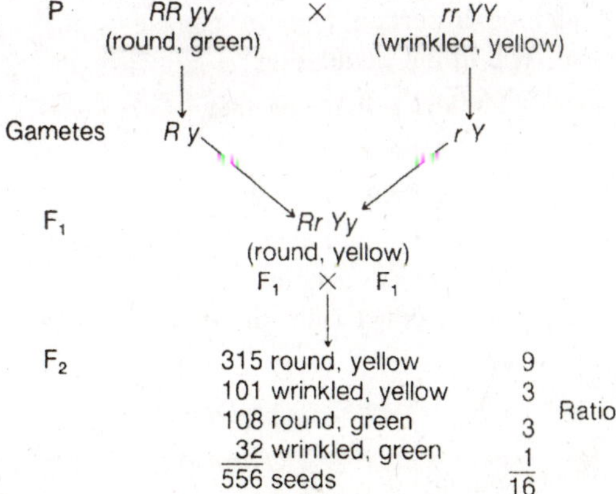

Fig. 8.4. The F_2 generation resulting from parents differing in two characters.

This is a convenient point to introduce some elementary rules of probability that we will often use throughout this book.

Rules of Probability

1. *Definition of probability.*

$$\text{Probability} = \frac{\text{The number of times an event is expected to happen}}{\text{The number of opportunities for it to happen}}$$

For example, the probability of rolling a four on a die in a single trial is written

$$p(\text{of a four}) = 1/6$$

because the die has six sides. If each side is equally likely to turn up, then the average result should be one four for each six trials.

2. *The product rule.* The probability of two independent events occurring simultaneously is the product of each of their respective probabilities. For example, with two dice we have independent objects, and

$$p(\text{of two fours}) = 1/6 \times 1/6 = 1/36$$

3. *The sum rule.* The probability of *either one* of two mutually exclusive events occurring is the sum of their individual probabilities. For example, with two dice,

$$p(\text{of two fours } \textit{or} \text{ two fives}) == 1/36 + 1/36 = 1/18$$

In the pea example, the F_2 of a dihybrid cross can be predicted if the mechanism for putting *R* or *r* into a gamete is *independent* of the mechanism for putting *Y* or *y* into a gamete. The frequency of gamete types can be calculated by determining their probabilities according to the rules just given. That is, if you pick a gamete at random, the *probability* of picking a certain type of gamete is the same as the frequency of that type in the population.

We know from Mendel's first law that

$$Y \text{ gametes} = y \text{ gametes} = 1/2$$
$$R \text{ gametes} = r \text{ gametes} = 1/2$$

For an *RrYy* plant, the probability that a gamete will be *R* and *Y* is written *p*(*R Y*); similarly, *p*(*Ry*) denotes the probability that a gamete will be *R* and *y*. By the product rule, therefore,

$$p(RY) = 1/2 \times 1/2 = 1/4$$
$$p(Ry) = 1/2 \times 1/2 = 1/4$$
$$p(ry) = 1/2 \times 1/2 = 1/4$$
$$p(rY) = 1/2 \times 1/2 = 1/4$$

Thus we can represent the F_2 generation by a giant grid named (after its inventor) a Punnett square.

The probability of 1/16 shown for each box in the square is derived by further application of the product rule. The assortment of alleles into the male gametes and that into the female gametes are independent events. Thus, for example, the probability (or frequency) of *RRYY* zygotes (combining an *RY* male gamete with an *RY* female gamete) will be 1/4 × 1/4 = 1/16. Grouping all the types, we find the 9:3:3:1 ratio (now not so mysterious) in all its beauty.

round, yellow	9/16 or 9
round, green	3/16 or 3
wrinkled, yellow	3/16 or 3
wrinkled, green	1/16 or 1

The concept of independence of the two systems (round or wrinkled versus yellow or green) is important. This concept of *independent assortment* has been generalized to give the statement now known as Mendel's second law.

- *Mendel's Second Law*. During gamete formation the segregation of one gene pair is independent of other gene pairs.

A note of warning; we shall see later that the phenomenon of gene linkage is an important exception to Mendel's second law.

Note how Mendel's counting' led to the discovery of such unexpected regularities as the 9:3:3:1 ratio, and how a few simple assumptions (such as equal segregation and independent assortment) can explain this ratio that initially seems so baffling. Although it was unappreciated at the time, Mendel's approach was to provide the key to an understanding of genetic mechanisms.

Of course, Mendel went on to test this second law. For example, he crossed an F_1 dihybrid *RrYy* with a double-homozygous recessive strain *rryy*. Such a cross involving a homozygous recessive is now known as a *testcross*. For his testcross, Mendel predicted that the dihybrid *RrYy* should produce the gametic types *RY, rY, Ry,* and *ry* in equal frequency—that is, as shown along one edge of the Punnett square, in the frequencies 1/4, 1/4, 1/4, and 1/4. On the other hand, because it is homozygous, the *rryy* plant should produce only one gamete type (*ry*), regardless of equal segregation or independent assortment. Thus the progeny phenotypes should be a direct reflection of the gametic types from the *RrYy* parent (because the *ry* contribution from the *rryy* parent does not alter the phenotype indicated by the other gamete).

♂ / ♀	*R Y* 1/4	*R y* 1/4	*r y* 1/4	*r Y* 1/4
R Y 1/4	*RR YY* 1/16	*RR Yy* 1/16	*Rr Yy* 1/16	*Rr YY* 1/16
R y 1/4	*RR Yy* 1/16	*RR yy* 1/16	*Rr yy* 1/16	*Rr Yy* 1/16
r y 1/4	*Rr Yy* 1/16	*Rr yy* 1/16	*rr yy* 1/16	*rr Yy* 1/16
r Y 1/4	*Rr YY* 1/16	*Rr Yy* 1/16	*rr Yy* 1/16	*rr YY* 1/16

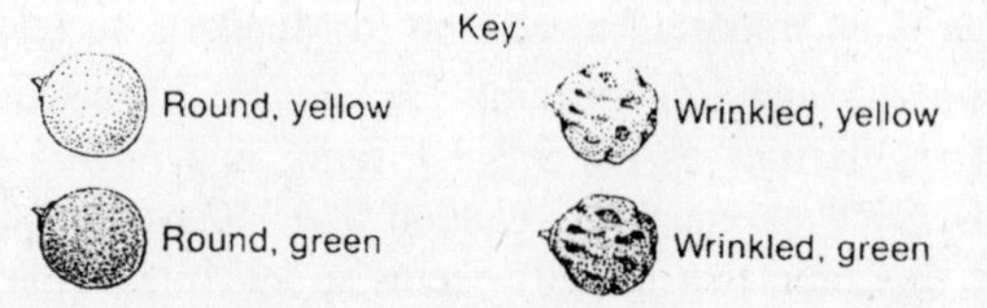

Fig. 8.5. Symbolic representation of the genetic and phenotypic constitution of the F_2 generation resulting from parents differing in two characters.

Mendel predicted a 1:1:1:1 ratio of *RrYy, rrYy, Rryy,* and *rryy* progeny from this testcross, and his prediction was confirmed. He tested the concept of independent assortment intensively on four different gene pairs and found that it applied to every combination.

Of course, the deduction of equal segregation and independent assortment as abstract concepts that explain the observed facts leads immediately to the question of what structures or forces are responsible for generating them. The idea of equal segregation seems to indicate that both alleles of a pair actually exist in some kind of orderly, paired configuration, from which they can separate cleanly during gamete formation. If any other gene pair behaves independently in the same way, then we have independent assortment. But this is all speculation at this stage of our discussion, as it was after the

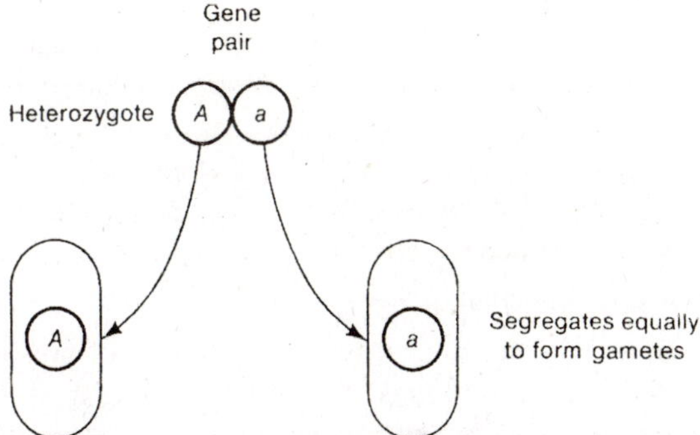

Fig. 8.6. Diagrammatic visualization of the equal segregation of one gene pair into gametes.

rediscovery of Mendel's work. The actual mechanisms are now known, and we shall discuss them later. (We shall see that it is the chromosomal location of genes that is responsible for then-equal segregation and independent assortment.) The key point to appreciate

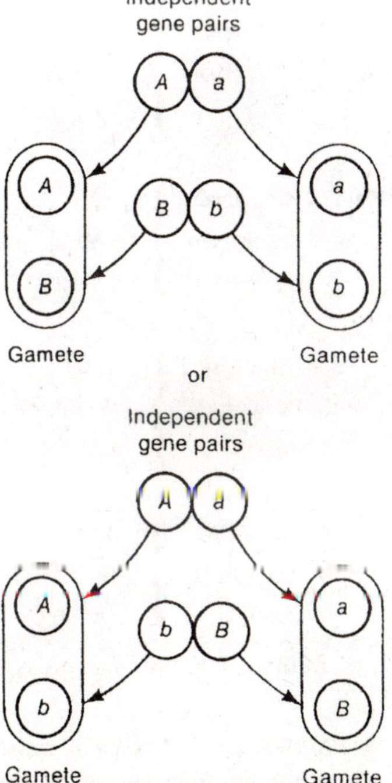

Fig. 8.7. Diagrammatic visualization of the segregation of two independent gene pairs into gametes.

at this stage is that the ground rules for genetic analysis were established by Mendel. His work made it possible to infer the existence and nature of hereditary particles and mechanisms without ever seeing them. All such theories were based on the analysis of phenotype frequencies in controlled crosses; this is the experimental approach still used in much of modern genetics.

Methods for Working Problems

We pause here for a few words on the working of problems. The Punnett square is sure and graphic, but it is unwieldy; it is suited only for illustration, not for efficient calculation.

A branch diagram is useful for solving some problems. For example, the 9:3:3:1 ratio can be derived by drawing a branch diagram and applying the product rule to determine the frequencies. (Note the use of the convention that *R*-represents both *RR* and *Rr*. That is, either allele can occupy the space indicated by the dash.)

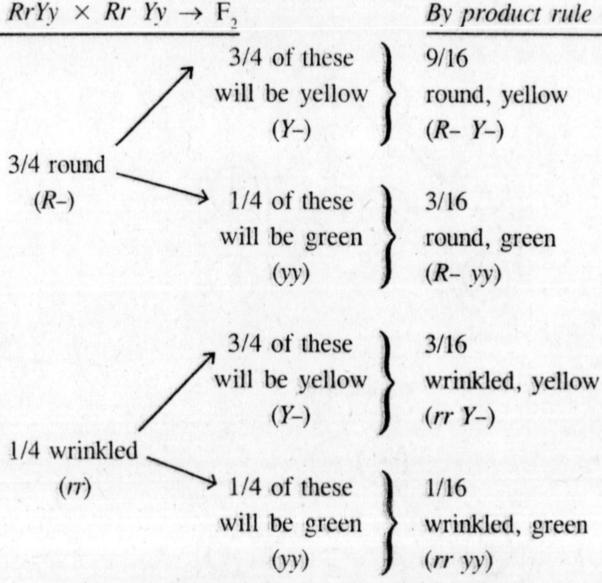

The branch diagram is, of course, a graphic expression of the product rule.

The diagram can be extended to a trihybrid ratio (such as *Aa Bb Cc* × *Aa Bb Cc)* by drawing another set of branches on the end. However, as the number of gene pairs increases, the number of identifiable phenotypes rises startlingly, and the number of genotypes climbs even more steeply. With such large class numbers, even the branch method becomes unwieldy.

Table 8.2. Rise in number of genotypic classes as the power of the number of segregating gene pairs

Number of segregating gene pairs	*Number of phenotypic classes*	*Number of genotypic classes*
1	2	3
2	4	9
3	8	27
4	16	81
.	.	.
.	.	.
.	.	.
n	2^n	3^n

In such cases, we must resort to devices based directly on the product and sum rules. For example, what proportion of progeny from the cross *Aa Bb Cc Dd Ee Ff* × *Aa Bb Cc Dd Ee Ff* will be *AA bb Cc DD ee Ff*? The answer is easily obtained if the gene pairs all assort independently, thereby allowing use of the product rule. That is, 1 /4 of the progeny will be *AA*, 1/4 will be *bb,* 1/2 will be *Cc,* 1/4 will be *DD,* 1/4 will be *ee,* and 1/2 will be *Ff,* so we obtain the answer by multiplying these frequencies:

$$p(AA\ bb\ Cc\ DD\ ee\ Ff) = 1/4 \times 1/4 \times 1/2 \times 1/4 \times 1/4 \times 1/2 = 1/1024$$

Let us return now to Mendel's work.

When Mendel's results were rediscovered in 1900, his principles were tested in a wide spectrum of eukaryotic organisms (those whose cells contain nuclei). The results of these tests showed that Mendelian genetics (sometimes with extensions that we shall discuss in the next three chapters) is universally applicable. Mendelian ratios (such as 3:1, 1:1, 9:3:3:1, and 1:1:1:1) were extensively reported, suggesting that equal segregation and independent assortment are fundamental hereditary processes found throughout nature. His laws are not merely laws about peas, but laws of the genetics of eukaryotic organisms in general. The experimental approach used by Mendel can be extensively applied in plants. However, in some plants, and in animals, the technique of selfing is impossible. This problem can be circumvented by intercrossing identical genotypes. For example, an F_1 animal resulting from the mating of parents from differing pure lines can be mated to its F_1 siblings (brothers or sisters) and an F_2 produced. The F_1 individuals are genetically identical, so the F_1 cross amounts to a selfing.

Simple Mendelian Genetics in Humans

Other systems present some special problems in the application of Mendelian methodology. One of the most difficult, yet most interesting, is the human species. Obviously, controlled crosses cannot be made, so human geneticists must resort to a scrutiny of established matings in the hope that informative matings have been made by chance. The scrutiny of established matings is called *pedigree analysis*. A member of a family who first comes to the attention of a geneticist is called the *propositus*. Usually the phenotype of the propositus is exceptional in some way—for example, a dwarf. The investigator then traces the history of the character shown to be interesting in the propositus back

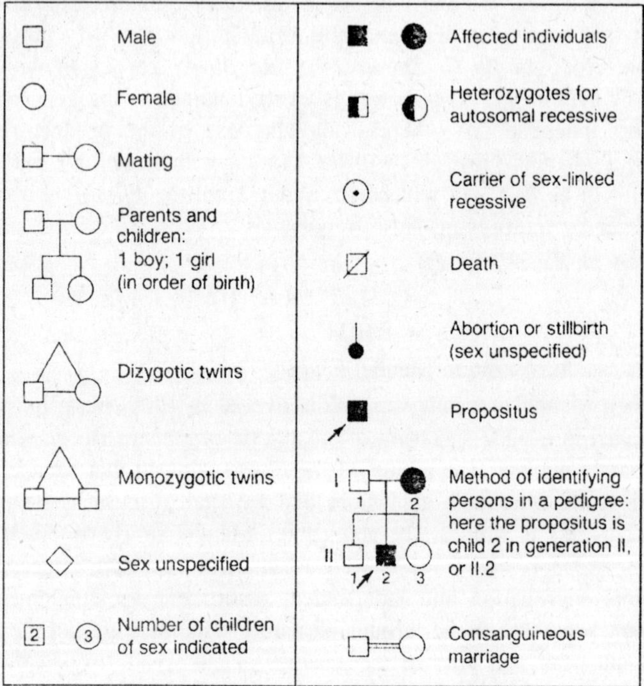

Fig. 8.8. Symbols used in human pedigree analysis.

through the history of the family, and a family tree or pedigree is drawn up using certain standard symbols.

Many human diseases and other exceptional conditions are determined by simple Mendelian recessive alleles. There are certain clues in the pedigree that must be sought. Characteristically the condition appears in progeny of unaffected parents. Furthermore, two affected individuals cannot have an unaffected child. Quite often such recessive alleles are revealed by consanguineous matings—for example, cousin marriages. This is particularly true of rare conditions where chance matings of heterozygotes are expected to be extremely rare. It has been estimated, for example, that first-cousin marriages account for about 18 to 24 percent of albino children and 27 to 53 percent of children with Tay-Sachs disease; both are rare recessive conditions. Some other examples of disease-causing recessive alleles in humans are those for cystic fibrosis and phenylketonuria (PKU). Of course, variants that are not regarded as diseases also may be caused by recessive alleles. These may be rare as in albinism or common as in light eye colour (blue or green) in North American populations.

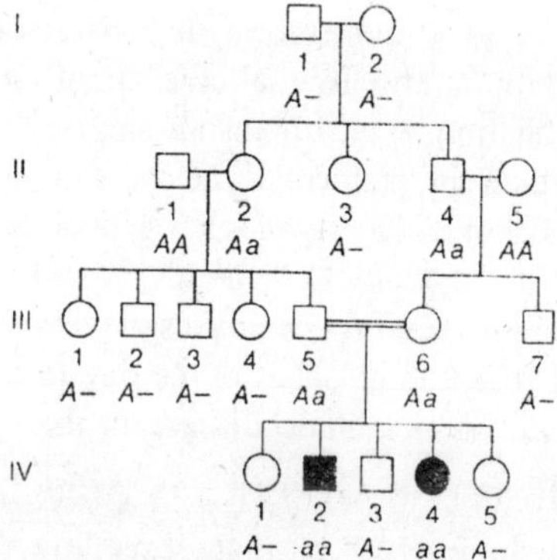

Fig. 8.9. Illustrative pedigree, involving an exceptional recessive phenotype determined by the recessive Mendelian allele a.

There are also examples of exceptional conditions caused by dominant alleles. (This is in contrast to the situation for such conditions as PKU, where the normal condition is attributable to the dominant allele, and the recessive allele causes PKU.) Once again, there are some simple rules to follow to discern from pedigrees a condition caused by a dominant allele: the condition typically occurs in every generation; unaffected individuals never transmit the condition to their offspring; two affected parents may have unaffected children; and the

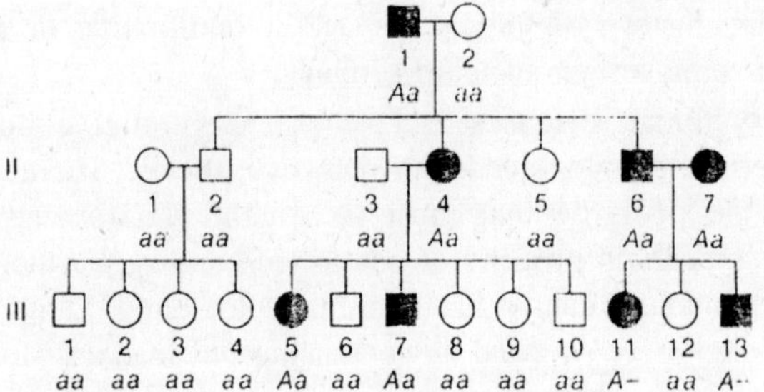

Fig. 8.10. Illustrative pedigree involving an exceptional dominant phenotype determined by the dominant Mendelian allele A.

condition is passed, on average, to one-half of the children of an affected individual. As with recessive alleles acting in a Mendelian manner, both sexes may be equally affected. Achondroplasia (a kind of dwarfism), Huntington's chorea, and brachydactyiy (very short fingers) are examples of exceptional conditions in humans caused by dominant alleles.

Notice that, in this kind of Mendelian analysis, there is again the notion of identification of genes affecting major biological function, this time in humans. Pedigrees for PKU, for example, demonstrate that there is a kind of gene controlling the character we might call "normal PKU function." The two alleles stand for presence and absence. This identification is an important step toward discovery of the precise way in which the abnormal allele is failing and its possible correction. The medical applications of such genetic analysis obviously are far-reaching. In fact, medical genetics is today a key part of medical training. Simple pedigree analyses have extensive use, not only in such medical research but also in the day-to-day counseling of prospective parents who fear genetic disease in their children.

Simple Mendelian Genetics in Agriculture

There has been an interest in plant breeding since prehistoric times. The methods used by Neolithic farmers were probably the same as those used until the discovery of Mendelian genetics. Basically the approach was to select superior phenotypes from seeds or plants derived from natural populations. Particularly desirable were pure lines of favourable phenotype, because these lines produced constant results over generations of planting. Without the knowledge of Mendelian genetics, how is it possible to develop pure lines? It so happens that self-pollinating plants, such as many crop plants, naturally lend lo be homozygous, and any heterozygous gene pairs become homozygous over generations of selfing. Thus pure lines have developed automatically over the years. However, these less sophisticated breeding practices suffered from a major problem: the breeder was forced to rely on favourable combinations of genes that occurred in nature. With the advent of Mendelian genetics, it became evident that favourable qualities in different lines could be combined through hybridization and subsequent gene reassortment. This procedure forms the basis of modern plant, breeding.

For naturally self-pollinating plants, such as rice or wheat, two pure lines (each of different favourable genotype) are hybridized by manual cross-pollination, and hence an F_1 is developed. The F_1 is then allowed to self, and its heterozygous gene pairs assort to produce many different genotypes, some of which represent desirable new combinations of the parental genes. A small proportion of these new genotypes will be pure-breeding already, but if not, several generations of selfing will produce homozygosity of the relevant genes.

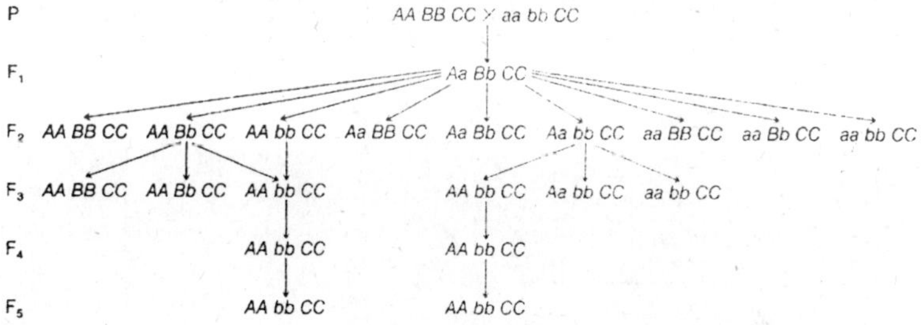

Fig. 8.11. The basic technique of plant breeding.

An example of genetic improvement in a species more familiar to most is the tomato. Anyone who has read a recent seed catalog will be familiar with the abbreviations V, F, and N next to a listed tomato variety. These represent, respectively, resistance to the pathogens *Verticillium*, *Fusarium*, and nematodes—resistance that has been crossed into the tomatoes, typically from wild forms of tomato, Another familiar phenomenon in tomatoes is determinate as opposed to indeterminate growth pattern. Determinate plants are bushier and more compact, and they do not need as much staking. Determinate growth is caused by a recessive allele *sp* (self-pruning), which has been crossed into modern varieties. Another useful allele is *u* (uniform ripening); this allele eliminates the green patch or shoulder around the stem on the ripe fruit.

Such examples could be listed for many pages. The point is that simple Mendelian genetics (as in this chapter) has provided agricultural

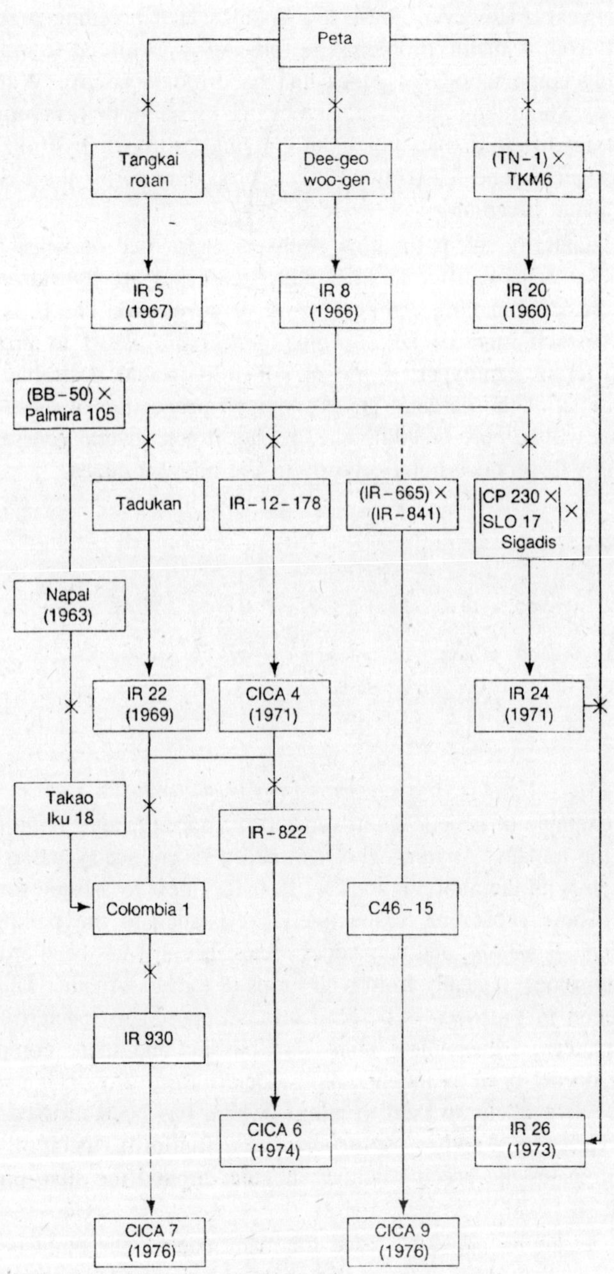

Fig. 8.12. The complex pedigree of modern rice varieties.

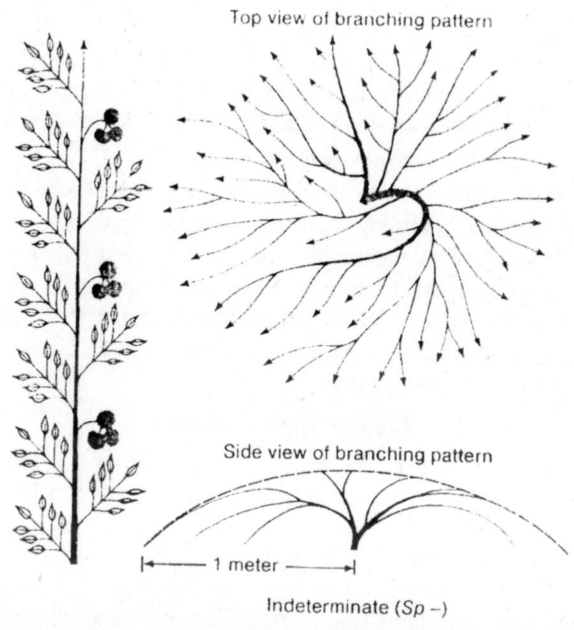

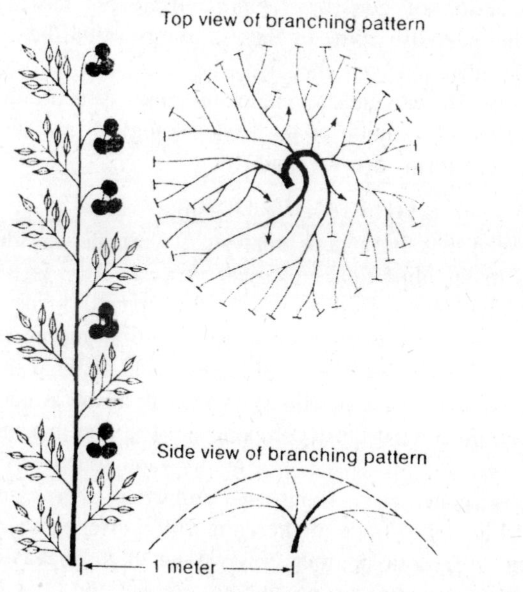

Fig. 8.13. Growth characteristics of indeterminate (Sp–) and determine (sp sp) tomatoes.

plant breeding with its rationale and its modern methods. Entire complex genotypes may be constructed from an array of ancestral lines, each showing some desirable feature. When we think of genetic engineering, we think of the genetic techniques of the 1970s and 1980s, but genetic engineering for plant improvement began long ago.

Mendelian genetics also has provided a formal theoretical basis for animal breeding, enabling a greater efficiency than under traditional practices. More recent techniques such as the use of frozen semen, artificial insemination, frozen embryos, and surrogate mothers in livestock species have enabled breeders to amplify the number of offspring of a specific genotype, a number normally limited by the lifespan of the maternal animal.

Origin of Variants

Genetic analysis, as we have seen, must start with parental differences. Without variants, no genetic analysis is possible. Where do these variants, this raw material for genetic analysis, come from? This is a question that can be answered in full only in later chapters. Briefly, most of the variants used by Mendel (and by ancestral and modern breeders of plants and animals) arise spontaneously, without the deliberate action of geneticists, in nature or in the breeders' populations. Undoubtedly, some of these variants would be "weeded out" by natural selection, but they may be kept alive through their deliberate nurture by geneticists, either for some specific breeding program or for the elucidation of the basic biological mechanisms of which the different forms are variants.

Genetic Dissection

We have seen that the genetic analysis of variants can identify a gene involved in an important biological process. This is a central aspect of modern genetics; this process is referred to as genetic dissection of biological systems. Mendel was the first genetic surgeon. Using genetic analysis, he was able to identify and distinguish among the several components of the hereditary process in a way as convincing as if he had microdissected those components. The fact that the genes he was using were for pea shape, pea colour, and so on was largely irrelevant. Those genes were being used simply as *genetic markers*, which enabled Mendel to trace the hereditary processes of segregation and assortment. A genetic marker is a variant allele that is used to label a biological structure or process throughout the course of an experiment. It is almost as though Mendel were able to "paint" two alleles different colours and send them through a cross to see how they behaved! Genetic markers are now routinely used in genetics and in all of biology to study all sorts of processes that the marker genes themselves do not directly affect.

Probably without realizing it, Mendel had also invented another aspect of genetic dissection in which the precise genes used *were* important. In this type of analysis, genetic variants are used in such a way that, by studying the variant gene function, we can make inferences about the "normal" operation of the gene and the process it controls. Let's consider the petal-colour gene in peas again. This is a major gene affecting an important biological function, which is the colour of the petals. Undoubtedly the success or failure of a plant in nature would largely hinge on its having the right kind of petal colour, presumably to attract the appropriate pollinator. So Mendel had pinpointed a gene for a process of central importance to the biology of peas, thus opening the way to extensive further study. Genetic dissection is a versatile tool of modern biological research.

Once a gene has been identified, affecting (say) petal colour in peas, we have called it a major gene, but what does this really mean? It is major in that it is obviously having a profound effect on the colour of the petals. But can we conclude that it is *the* single most important step in the determination of petal colour? The answer is no, and the reason may be seen in an analogy. If we were trying to discover how a car engine works, we might investigate this by pulling out various parts and observing the effect on the running of the engine. If a battery cable were disconnected, the engine would stop; we might erroneously conclude that this cable is the most important part of the running of the engine. Obviously, other parts are equally necessary, and their removal could also stop or seriously cripple the engine. In a similar way it can be shown that several genes can be identified, all of which have a major and similar effect on petal colouration.

Mendel's work has withstood the test of time and has provided us with the basic groundwork for all modern genetic study. Yet his work went unrecognized and neglected for 35 years following its publication. Why? There are many possible reasons, but here we shall consider just one. Perhaps it was because biological science at that time could not provide evidence for any real physical units within cells that might correspond to Mendel's genetic particles. Chromosomes had certainly not yet been studied, meiosis not yet described, and even the full details of plant life cycles had not been worked out. Without this basic knowledge, it may have seemed that Mendel's ideas were mere numerology.

9

Extensions to Mendelian Analysis

We have seen that Mendel's laws seem to hold across the entire spectrum of eukaryotic organisms—that is, we can identify analogous phenomena that reveal segregation and independent assortment. These laws form a base for predicting the outcome of simple crosses. However, it is only a base; the real world of genes and chromosomes is more complex than is revealed by Mendel's laws, and exceptions and extensions abound. These situations do not invalidate Mendel's laws. Rather, they show that there are more situations than can be explained by segregation and independent assortment of gene pairs and that these situations must be accommodated into the fabric of genetic analysis. This is the challenge we now must meet. Of course, one extension has already been accommodated—sex linkage. This chapter presents a grab bag of other extensions, and the two following chapters discuss two major extensions. We shall see that, rather than creating a hopeless and bewildering situation, these complexities combine to form a precise and unifying set of principles for the genetic analyst. These principles interlock and support each other in a highly satisfying way that has provided great insight into the mechanics of inheritance.

Variations on Dominance Relations

Dominance is a good place to start. Mendel observed (or at least reported) full dominance (and recessiveness) for all the seven gene pairs he studied. He may have been selective in his choice of pea characters to study, because variations on the basic theme crop up quite often in analysis. The problems center on the phenotype of the heterozygote. Some examples will illustrate this. In four-o'clock plants, when a pure line with red petals is crossed to a pure line with white petals, the F_1 have *not* red petals but pink! If an F_2 is produced, the result is

1/4 red petals	1 C_1C_1
1/2 pink petals	2 C_1C_2
1/4 white petals	1 C_2C_2

The occurrence of an intermediate phenotype in the heterozygote introduces the possibility of *incomplete dominance*. The precise position of the heterozygote on the phenotypic "scale" defines several possibilities. In practice it is often difficult to determine exactly where on the scale the heterozygote is, however.

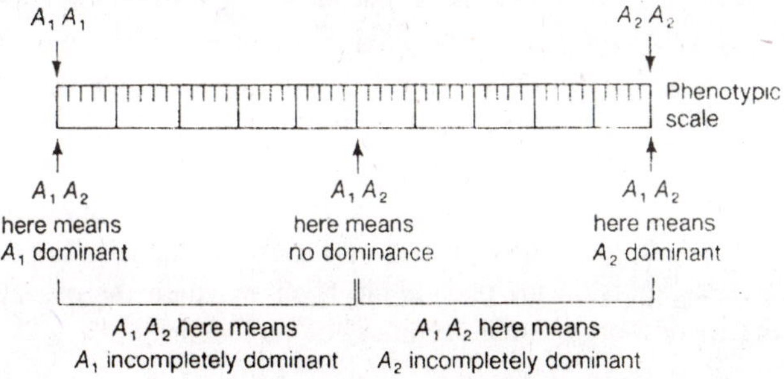

Fig. 9.1. Summary of dominance relationships.

The phenotype of the heterozygote is also the key in the phenomenon of *codominance*, in which the heterozygote shows the phenotypes of *both* the homozygotes. In a sense, then, codominance is no dominance at all! A good example is found in the M-N blood-group gene pair in humans. Three blood groups are possible—M, N, and MN—and these are determined by the genotypes L^ML^M, L^NL^N, and L^ML^N, respectively. Blood groups actually represent the presence of an immunological antigen on the surface of red blood cells. People of genotype L^ML^N have both antigens. This example shows how the heterozygote can show both phenotypes. Another interesting example of dominance is found in the human disease sickle-cell anemia. The gene pair concerned affects the oxygen transport molecule hemoglobin, the major constituent of red blood cells. The three genotypes have different phenotypes, as follows:

Hb^AHb^A: Normal. Red blood cells never sickled.

Hb^SHb^S: Severe, often fatal anemia. Red blood cells sickle-shaped.

Hb^AHb^S: No anemia. Red blood cells sickle only under abnormally low oxygen concentrations.

In regard to anemia the Hb^1 allele is dominant. In regard to blood cell shape there is incomplete dominance. Penally, as we shall now see, in regard to hemoglobin there is codominance! Luckily, the two types of hemoglobin concerned, dictated by the two alleles, have different chemical charges, and this property can be used in a technique called electrophoresis to make a more refined distinction.

In electrophoresis, mixtures of proteins can be separated on the basis of their charges. A small sample of protein (here, hemoglobin from red blood cells) is placed in a small well, cut in a slab of gel. A powerful electric field is applied across this gel, and the hemoglobin moves according to the degree of electrostatic charge. The gel is actually a supporting web containing electrolyte solution in its spaces: hence the hemoglobin can move easily through the field. After an appropriate time, the gel can be stained for protein. The blotches are the stained hemoglobin. We see that homozygous normal people have one type of hemoglobin (A), and anemics have another slow-moving type (S). The heterozygotes have both types, A and S.

Sickle-cell anemia illustrates the somewhat arbitrary nature of the terms incomplete dominance and codominance. The type of dominance depends on the phenotypic level at which the observations are being made—organismal, cellular, or molecular.

Multiple Alleles

Early in the history of genetics, it became clear that it is possible to have more than two forms of one kind of gene. Although only two actual alleles of a gene can exist in a diploid cell (and only one in a haploid cell), the total number of possible different allelic forms that might exist in a population of individuals is often quite large. This situation is called *multiple allelism*, and the set of alleles itself is called an *allelic series*. The concept of allelism is a crucial one in genetics, so we consider several examples. The examples themselves serve also to introduce important areas of genetic investigation.

ABO Blood Group in Humans

The human ABO blood-group alleles afford a modest example of multiple allelism. There are four blood types (or phenotypes) in the ABO system. The allelic series includes three major alleles, which can be present in any pairwise combination in one individual; thus only two of the three alleles can be present in any one individual. Note that the series include cases of both complete dominance and codominance. In this allelic series, the alleles I^A and I^B each determine a unique form of one type of antigen; the allele *i* determines a failure to produce either form of that type of antigen.

Table 9.1. ABO blood groups in human

Blood phenotype	*Genotype*
O	ii
A	I^AI^A or I^Ai
B	I^BI^B or I^Bi
AB	I^AI^B

C Gene in Rabbits

Another example of multiple allelism involves a larger allelic series determining coat colour in rabbits. The alleles in this series are C (full colour), c^{ch} (chinchilla, a light grayish colour), c^h (Himalayan, albino with black extremities), and c (albino). You will have noted the use of a superscript to indicate an allele of a type of gene; this symbolism often is necessary because more than the two symbols C and c are needed to describe multiple alleles. In this series, dominance is in the order of the alleles.

Table 9.2. C gene in rabbits

Coat colour phenotype	*Genotype*
Full colour	CC or Cc^{ch} or Cc^h or Cc
Chinchilla	$c^{ch}c^{ch}$ or $c^{ch}c^h$ or $c^{ch}c$
Himalayan	c^hc^h or c^hc
Albino	cc

Operational Test of Allelism

Now that we have seen two examples of allelic series, it is a good time to pause and consider a question. How do we know that a set of phenotypes is determined by alleles of the same type of gene? In other words, what is the operational test for allelism? For now, the answer is simply the observation of Mendelian single-gene-pair ratios in all combinations crossed. For example, consider three pure-line phenotypes in a hypothetical plant species. Line 1 has round spots on the petals; line 2 has oval spots on the petals; and line 3 has no spots on the petals. Suppose that crosses of the three lines yield the following results:

These results prove that we are dealing with three alleles of a single gene that affects petal spotting. We can choose any symbols we

Cross	F_1	F_2
1 × 2	all round-spotted	3/4 round, 1/4 oval
1 × 3	all round-spotted	3/4 round, 1/4 unspotted
2 × 3	all oval-spotted	3/4 oval, 1/4 unspotted

wish. Because we don't know which phenotype is the wild-type, we could follow the rabbit system and use *S* for the round-spotted allele, s° for the oval-spotted allele, and *s* for the unspotted allele. Alternatively, we could use S^r for round, S° for oval, and *s* for unspotted. The convention provides no firm rules about whether to use capital or small letters, particularly for the alleles in the middle of the series that are dominant to some of their alleles but recessive to others.

Clover Chevrons

Clover is the common name for plants of the genus *Trifolium*. There are many species—some native to North America and some that grow here as introduced weeds. The red-flowered and white-flowered clovers are familiar to most people. These are *different* species, but each shows considerable variation among individuals in the curious V or "chevron" pattern on the leaves. Much genetic research has been done with white clover; the different chevron forms (and the absence of chevrons) constitute an allelic series in this species. Furthermore, several types of dominance relations may be seen in the allele combinations.

Incompatibility Alleles in Plants

It has been known for millennia that some plants just will not self-pollinate. A single plant may produce both male and female gametes, but no seeds will ever be produced. The same plants, however, will cross with certain other plants, so obviously they are not sterile. This phenomenon is called *self-incompatibility*. Of course, the pea plants used by Mendel were not self-incompatible, for he was able to self his plants with ease. We now know that incompatibility has a genetic basis and that there are several different genetic systems acting in different self-incompatible species. These systems form very nice examples of multiple allelism. One of the most common systems is found in many dicot and monocot plants, including sweet cherries, tobacco, petunias, and evening primroses. In each of these species, one gene determines compatibility/incompatibility relations, with many different allelic forms of the gene possible in different plants of any one species.

If a pollen grain bears an S allele that is also present in the maternal parent, then it will not grow; however, if that allele is not in the maternal tissue, then the pollen grain produces a pollen tube containing the male nucleus, and this tube effects fertilization. The number of *S* alleles in a series in one species can be very large (over 50 in the evening primrose and clover), and cases of more than 100 alleles have been reported in some species.

Tissue Incompatibility in Humans

When an organ or a tissue graft is medically necessary in a human, the success or failure of the graft depends on the genotypes of the host and the donor. If the two are mismatched, rejection of the graft eventually occurs, often accompanied by death of the host.

The rejection system is based on two important genes called *HLA-A* and *HLA-B,* which determine the immunological acceptability of the introduced tissue. Each gene has a series of allelic forms. There are eight alleles for *HLA-A,* designated *A1, A2, A3, A9, A10, A11, A28,* and *A29.* By coincidence, there are also eight alleles for *HLA-B,* designated *B5, B7, B8, B12, B13, B14, B18,* and *B27.* (The two genes do not show independent assortment, but this need not concern us now.

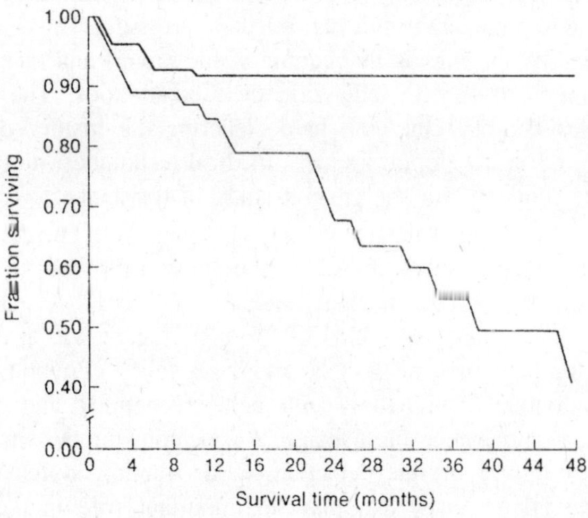

Fig. 9.2. The success of kidney transplants in HLA-matched and HLA-mismatched sibling pairs.

For the moment we shall simply use these allelic series as excellent examples of multiple allelism in humans.)

The HLA type of an individual is determined by testing his or her lymphocytes (white blood cells) with a battery of standard antibodies from donors of known HLA type. In a transplant, the recipient's immune system will recognize the graft as "foreign" and reject it if an allele is present in the donor tissue that is not present in the recipient. In the typing test, this rejection reaction is observed as lymphocyte death. The immune system will not reject tissue that lacks some alleles present in the recipient—it only rejects "strange" alleles.

Table 9.3. Success or failure of transplants determined by HLA types

Transplant	*Recipient genotype*	*Donor genotype*	*Result*
1	*A1 A2, B5 B5*	*A1 A1, B5 B7*	Rejected (because of *B7*)
2	*A2 A3, B7 B12*	*A1 A2, B7 B7*	Rejected (because of *A1*)
3	*A1 A2, B7 B5*	*A1 A2, B7 B7*	Accepted
4	*A2 A3, B7 B5*	*A3 A3, B5 B5*	Accepted

We might wonder what function these HLA genes normally serve—obviously, tissue transplants have not been a normal aspect of human evolution! One possible answer is that some HLA alleles seem to confer resistance or susceptibility to certain specific diseases. Thus these alleles may have played some role in our evolutionary history, in response to some environmental selection pressure. This supposition is supported by the finding that certain ethnic groups and races show a preponderance of specific alleles in their populations. The study of such alleles is important both in deciphering the history of human evolution and in developing modern medical techniques—not only for transplants, but also for the general study of resistance to disease.

Another very interesting possibility emerging from current research is that the HLA genes may be intimately involved in cancer. Presumably, one of the normal functions of the HLA genes is to recognize cancer cells as a kind of "foreign" agent within the body. The development of cancer cells may be a fairly common event in healthy individuals, with these cells being recognized and destroyed before they cause noticeable damage. A cancerous tumour- may be the result of a failure of this detection-and-destruction system at some stage. The HLA genes also play an important role in the normal immune response.

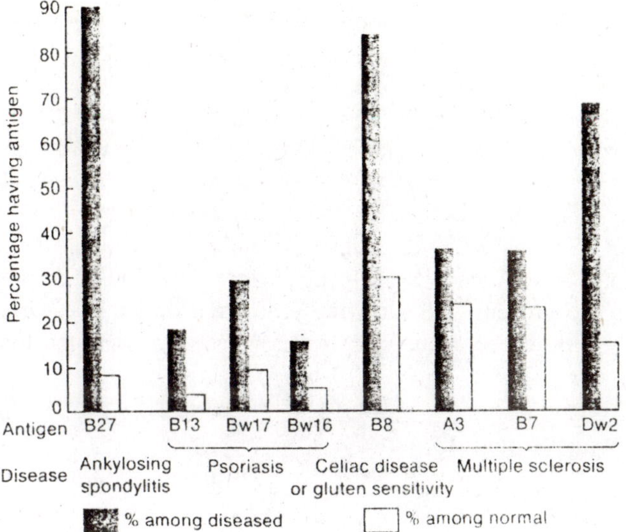

Fig. 9.3. Histograms showing effects of specific HLA alleles on predisposition to certain diseases in humans.

As a postscript to this section, it is worth noting that several different "kinds" or "species" of cell-surface anti-genes have been discovered. Each of these kinds is, of course, specified by one kind of gene. The different forms one kind of antigen can take are determined by the multiple alleles of the gene concerned.

Lethal Genes

Normal wild-type mice have coats with a rather dark overall pigmentation. In 1904, Lucien Cuenot studied mice having a lighter coat colour called yellow. Mating a yellow mouse to a normal mouse from a pure line, Cuenot observed a 1:1 ratio of yellow to normal mice in the progeny. This observation suggests that a single gene determines this aspect of coat colour and that an allele for yellow is dominant to an allele for normal colour. However, the situation became more confusing when he crossed yellow mice with one another. The result was always the same, no matter which yellow mice were used:

$$\text{yellow} \times \text{yellow} \begin{cases} \nearrow \ 2/3 \text{ yellow} \\ \searrow \ 1/3 \text{ normal colour} \end{cases}$$

Two things are of interest in these results. First, the 2:1 ratio is a departure from Mendelian expectations. Second, Cuenot failed to detect (in any generation of breeding) a homozygous yellow mouse.

Cuenot suggested the following explanation for these results. A cross between two heterozygotes would be expected to yield a Mendelian genotype ratio of 1:2:1. If one of the homozygous classes died before birth, the live births would then show a 2:1 ratio of heterozygotes to the surviving homozygotes. In other words, the allele A^Y for yellow is dominant to the normal allele A with respect to its effect on colour, and it also acts as a *recessive lethal* allele with respect to a character we could call "survival." Thus a mouse with the homozygous genotype A^YA^Y dies before birth and is not observed among the progeny. All surviving yellow mice must be heterozygous A^YA, so a cross between yellow mice will always yield the following results:

	↗	1/4 AA	normal
$A^YA \times A^YA$	→	2/4 A^YA	yellow
	↘	1/4 A^YA^Y	die before birth

The expected Mendelian ratio of 1:2:1 would be observed among the zygotes, but it is altered to a 2:1 ratio of viable progeny because of the lethality of the A^YA^Y genotype. This hypothesis was confirmed by the removal of uteri from pregnant females of the yellow × yellow cross; one-fourth of the embryos were found to be dead.

The A^Y allele has effects on two characters: coat colour and survival. Such genes known to have more than one distinct phenotypic effect are called *pleiotropic* genes. It is entirely possible that both effects of the A^Y pleiotropic allele are the result of the same basic cause, which promotes yellowness of coat in a single dose and death in a double dose.

Lethal genes are quite common, even in humans. Some lethal effects occur in utero, others much later—in infancy, in childhood, or even in adulthood. Only rarely is a lethal gene associated with a distinguishable heterozygous phenotype, as in yellow mice. One other example you may be familiar with is the tail-less condition in Manx cats. The determining gene is homozygous lethal. Typically, the only detectable effect of a lethal gene is on survival of the individual, In some cases, a specific abnormality can be pin-pointed as the cause of death. For example, a recessive lethal gene in rats may have its pleiotropic effects traced to a basic problem in the nature of the rat's cartilage.

The lethality of an allele of a gene is often dependent on the environment in which the organism develops. Whereas certain alleles would be lethal in virtually any environment, others are viable in one

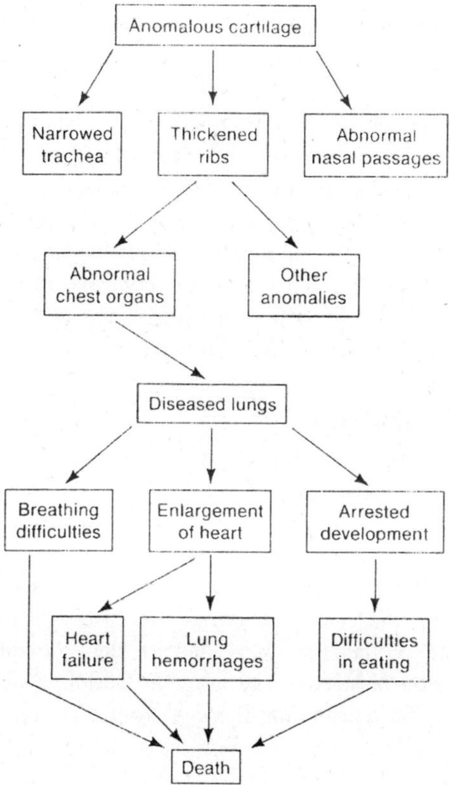

Fig. 9.4. Diagram showing how one specific lethal gene causes death in rats.

environment but lethal in another. For example, remember that many of the phenotypes favoured and selected by agricultural breeders would almost certainly be wiped out in nature in competition with the normal members of their population. Modern grain varieties provide good examples; it is only the careful nurturing by the farmer that has maintained such phenotypes for our benefit.

In practical genetics, we very commonly encounter situations where perfect expected Mendelian ratios are consistently skewed in one direction by one allele. For example, in the cross $Aa \times aa$, we predict a progeny phenotypic ratio of 50 percent $A-$ and 50 percent aa, but we might consistently observe a ratio such as 55%:45% or 60%:40%. In such a case, the a allele is said to be *subvital*—a kind of partial lethality, expressed in only some individuals. The partial lethality may range from 0 to 100 percent depending on the rest of the genome and on the environment.

Several Genes Affecting the Same Character

We saw earlier that, in a genetic dissection, the identification of a major gene affecting a character does not mean that this is the *only* gene affecting that character. An organism is a highly complex machine in which all functions interact to a greater or lesser degree. At the level of genetic determination, the genes likewise can be regarded as interacting. A gene does not act in isolation; its effects depend not only on its own functions but also on the functions of other genes (and on the environment). In many cases, the complex interactions of major genes are detectable through genetic analysis, and we now look at some examples. Typically, the mode of interaction is revealed by *modified Mendelian ratios*.

Coat Colour in Mammals

A character that has been extensively studied at the genetic level is coat colour in mammals. These studies have revealed a beautiful set of examples of the interplay between different genes in the determination of one character. The best studied mammal in this regard is the mouse, because of its small size and its short reproductive cycle. However, the genetic determination of coat colour in mice has direct parallels in other mammals, and we shall look at some of these as our discussion proceeds. There are at least five interacting major genes involved in determining the coat colour of mice: *A, B, C, D,* and S.

A gene

The wild-type allele *A* produces a phenotype called agouti. Agouti is an overall grayish colour with a "brindled" or "mousy" appearance. It is common in many mammals in nature. The effect is produced by a band of yellow on the hair shaft. In the nonagouti phenotype (determined by the allele *a*), the yellow band is absent, so the coat colour appears solid.

The lethal yellow A^Y allele is another form of this gene. Another form is a^t, which gives a "black-and-tan" effect involving a cream-coloured belly with dark pigmentation elsewhere. We shall not include these two alleles in the following discussion.

B gene

There are two major alleles of the *B* gene. The allele *B* gives the normal agouti colour in combination with *A*, but with *aa* it gives solid black. The genotype *A-bb* gives a colour called cinnamon ("mousy" brown), and *aa bb* gives solid brown.

The following sample cross illustrates the inheritance pattern of the *A* and *B* genes.

AA bb (cinnamon) × *aa BB* (black)
or *AA BB* (agouti) × *aa bb* (brown)

↓

F_1 all *Aa Bb* (agouti)

↓

F_2 9 *A–B*–(agouti)
3 *A–bb* (cinnamon)
3 *aa B* – (black)
1 *aa bb* (brown)

In horses, no *A* agouti gene seems to have survived the generations of breeding, although such a gene does exist in certain wild relatives of the horse. The colour we have called brown in mice is called chestnut in horses, and this phenotype also is recessive to black.

***C* gene**

The wild-type allele *C* permits colour expression, and the allele *c* prevents colour expression. The *cc* constitution is said to be *epistatic* to the other colour genes. The word epistatic literally means "standing upon"; the *c* allele in homozygous condition "stands on" (blots out) the expression of other genes concerned with coat colour. The *cc* animals, lacking coat colour, are called albino. Albinos are common in many mammalian species, but albinos have also been occasionally reported among birds, snakes, and fish. Epistatic genes produce interesting modified ratios, as seen in the following sample cross (where we assume that both parents are *aa*):

BB cc (albino) × *bb CC* (brown)
or *BB CC* (black) × *bb cc* (albino)

↓

F_1 all *Bb Cc* (black)

↓

F_2 9 *B-C*-(black) 9
3 *bb C*-(brown) 3
3 *B-cc* (albino) } 4
1 *bb cc* (albino) }

A phenotypic ratio of 9:3 :4 is observed. This ratio is the signal for inferring gene interaction of the type called recessive epistasis. In some other organisms, dominant epistasis is observed. (What ratio is produced in dominant epistasis?)

We have already encountered the c^h (Himalayan) allele in rabbits. It exists also in other mammals, including mice (also called Himalayan) and cats (called Siamese).

It should be pointed out here that the term epistasis is often used in a different way (mainly in population genetics) to describe *any* kind of gene interaction.

D gene

The *D* gene controls the intensity of pigment specified by the other coat-colour genes. The genotypes *DD* and *Dd* permit full expression of colour in mice, but *dd* "dilutes" the pigment to a milky appearance. Dilute agouti, dilute cinnamon, dilute brown, and dilute black coats all are possible. A gene of this nature is called a *modifier gene*. In the following sample cross, we assume that both parents are *aa CC*.

BBdd (dilute black) × *bbDD* (brown)
or *BB DD* (black) × *bbdd* (dilute brown)
↓
F_1 all *Bb Dd* (black)
↓
F_2 9 *B-D-*(black)
3 *B-dd* (dilute black)
3 *bb D-*(brown)
1 *bb dd* (dilute brown)

In horses, the *D* allele shows incomplete dominance.

S gene

The *S* gene controls the presence or absence of spots. The genotype S-results in no spots, and *ss* produces a spotting pattern called piebald in both mice and horses. This pattern can be superimposed on any of the coat colours discussed earlier—with the exception of albino, of course.

By this time, the point of the discussion should be obvious. Normal coat appearance in wild mice is produced by a complex set of interacting genes determining pigment type, pigment distribution in the individual hairs, pigment distribution on the animal's body, and the presence or absence of pigment. Similar situations exist for any character in any organism.

Examples of Gene Interaction in Other Organisms

Some other kinds of gene interaction are best illustrated in other organisms. Peas provide a good example of one important situation. Two different, independently obtained pure lines of pea plants are both white-petaled. When these lines are crossed, all of the F_1 have purple flowers. The F_2 shows both purple and white plants in a ratio of 9:7. How can this result be explained? By now, you should immediately suspect that the 9:7 ratio is a modification of the Mendelian 9:3:3:1 ratio. The explanation is that two different genes in the pea have similar effects on petal colour. Let us represent the alleles of these genes by *A, a, B,* and *b*.

white strain 1 × white strain 2

AA bb × *aa BB*

↓

F_1 all *Aa Bb* (purple)

↓

F_2

9 *A-B-*(purple)	9
3 *A-bb* (white)	
3 *aa B-*(white)	7
1 *aa bb* (white)	

Both gene pairs affect petal colour. Whiteness can be produced by a recessive allele of either gene pair, but purpleness is a phenotype produced by a combination of the dominant alleles of *both* gene pairs. This phenomenon is called *complementary gene action*, a term that satisfactorily describes how the two dominant alleles are uniting to produce a specific phenotype—in this example, purple pigment. (Note that we might have mistakenly inferred purpleness to be a specific phenotype of one gene pair if we had only one white strain available for our crosses.)

Another important kind of interaction is *suppression* of one gene by another. This interaction is illustrated by the inheritance of the production of a chemical called malvidin in the plant genus *Primula*. Malvidin production is determined by a single dominant gene *K*. However, the action of this dominant gene may be suppressed by a nonallelic dominant suppressor *D*. The following pedigree is informative:

Recessive suppression of both dominant and recessive genes also is known. The suppressor gene may have its own associated phenotype or (as in the malvidin example) have no known phenotypic effect other than the suppression.

KK dd (malvidin) × *kk DD* (no malvidin)

↓

F_1 all *Kk Dd* (no malvidin)

↓

F_2 9 K-D-(no malvidin) } 13
3 *kk D*-(no malvidin)
1 *kk dd* (no malvidin)
3 *K-dd* (malvidin) 3

Our final example of gene interaction introduces a concept that we shall develop in a later chapter. It concerns the genes that control fruit shape in the plant called shepherd's purse. Two different lines have fruits of different shapes: one is "round," the other "narrow." Are these two phenotypes determined by two alleles of a single gene? A cross between the two lines reveals an F_1 with round fruit; this result is consistent with the hypothesis of a single gene pair. However, the F_2 shows a 15:1 ratio of round to narrow. Again, this ratio immediately suggests a modification of the 9:3:3:1 Mendelian ratio, and it can be explained in terms of two *duplicate genes*. Apparently, round fruits are produced as a result of the presence of at least one dominant allele of either gene. The two genes appear to be identical

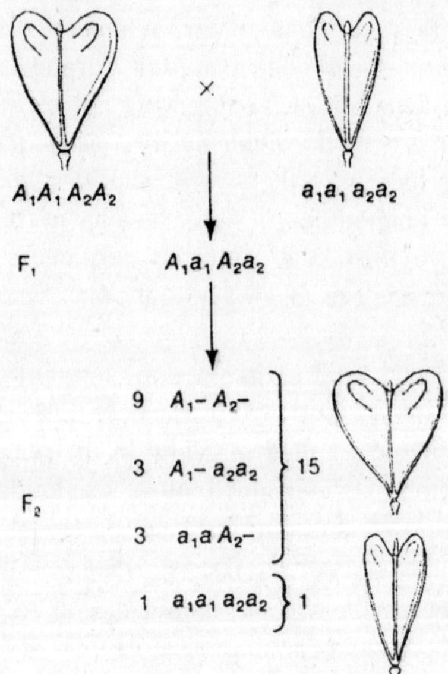

Fig. 9.5. Inheritance pattern of duplicate genes controlling fruit shape in shepherd's purse.

in function. (Contrast this 15:1 ratio with the 9:7 ratio obtained from complementary genes, where *both* dominant genes are necessary to produce a specific phenotype.)

Gene interactions also can be detected in haploid organisms. We have already discussed a gene in the fungus *Neurospora* where one allele causes albino asexual spores, as opposed to the normal pinkish-orange colour produced by the wild-type allele of the same gene. A cross *al* × *al*$^+$ gives 1/2 *al* and 1/2 *al*$^+$ progeny. Another interesting gene, *ylo,* gives yellow asexual spores, and the cross *ylo* × *ylo*$^+$ gives 1/2 *ylo* and 1/2 *ylo*$^+$ progeny. When an *al* culture is crossed with a *ylo* culture, the resulting progeny are 1/4 yellow, 1/4 wild-type, and 1/2 albino. How can we explain this result? The answer is a kind of epistasis: *al and ylo* are forms of separate genes, each of which can affect the normal production of pink pigment. The genotypes in the cross are the following:

Haploid parental culture	*al ylo*$^+$ (albino) × *al*$^+$ *ylo* (yellow)	
	↓	
Transient diploid	*al*$^+$/*al*, *ylo*$^+$/*ylo*	
	↓ meiosis	
Progeny (culture from sexual spores)	1/4 *al ylo* (albino)	} 1/4
	1/4 *al ylo*$^+$ (albino)	}
	1/4 *al*$^+$ *ylo* (yellow)	1/4
	1/4 *al*$^+$ *ylo*$^+$ (normal)	1/4

Penetrance and Expressivity

Clearly, genes do not act in isolation. A gene does *not* determine a phenotype by acting alone; it does so only in conjunction with other genes and with the environment. Although geneticists do routinely ascribe a particular phenotype to an allele of a gene they have identified, we must remember that this is merely a convenient kind of jargon designed to facilitate genetic analysis. This jargon arises from the ability of geneticists to isolate individual components of a biological process and to study them as part of genetic dissection. Although this logical isolation is an essential aspect of genetics, the message of this chapter is that a gene cannot act by itself.

In the preceding examples, the genetic basis of the dependence of one gene on another has been worked out. In other situations, where the phenotype ascribed to a gene is known to be dependent on other factors but the precise nature of those factors has not been established, the terms *penetrance* or *expressivity* may be very useful in describing the situation. We have already encountered penetrance in the discussion of lethal alleles. *Penetrance* is defined as the percentage of individuals with a given genotype who exhibit the phenotype associated with that genotype. For example, an organism may be of genotype *aa* or A– but may not express the phenotype normally associated with its genotype—because of the presence of modifiers, epistatic genes, or suppressors in the rest of the genome or because of a modifying effect of the environment. Penetrance can be used to describe such an effect when the exact cause is not known.

Expressivity, on the other hand, describes the degree or extent to which a given genotype is expressed phenotypically in an individual. Again, the lack of full expression may be due to the rest of the genome or to environmental factors. Obviously, both penetrance and expressivity variation are integral components of the concept of norm of reaction.

Human pedigree analysis and predictions in genetic counseling can often be thwarted by the phenomena of penetrance and expressivity. For example, if a disease-causing allele is not fully penetrant (as usually is the case), it is difficult to give a clean genetic bill of health to any individual involved as part of a disease pedigree—for example, individual R. On the other hand, pedigree analysis can sometimes identify individuals who do not express but almost certainly have a disease genotype—for example, individual Q. The exceptions to Mendelian analysis covered in this chapter and following chapters are

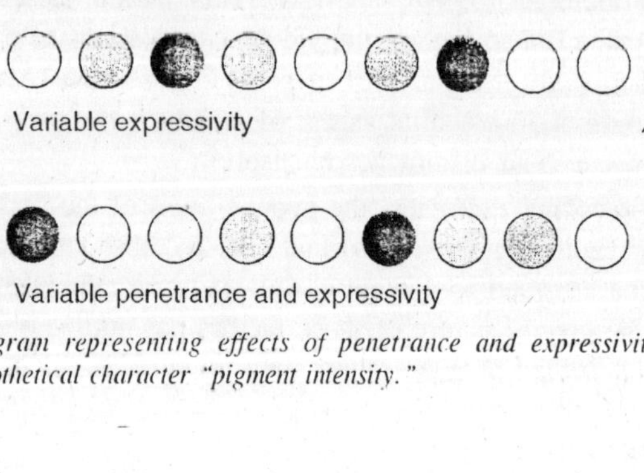

Phenotypic expression
(each circle represents an individual)

Variable penetrance

Variable expressivity

Variable penetrance and expressivity

Fig. 9.6. Diagram representing effects of penetrance and expressivity through a hypothetical character "pigment intensity."

not the only ones encountered in routine genetic analysis; they are simply among the most common. Each exception—whether, for example, multiple allelism, incomplete dominance, epistasis, or variable expressivity—has its key recognition features at the experimental level. The analyst, must be constantly on the lockout for these and any other results that might indicate the uniqueness of any given situation. Such signals often lead to the discovery of new phenomena and to the opening up of new research areas. Although it has been shown that Mendel's laws apply to all eukaryotic organisms, these laws are only a base for understanding heredity. The real world of genes and chromosomes is much more complex. In addition to the full dominance that Mendel observed in his experiments, incomplete dominance and codominance may exist. In incomplete dominance, the phenotype of a heterozygote is intermediate between those of the homozygotes. In codominance, the heterozygote shows the phenotypes of both homozygotes.

In his experiments, Mendel reported genes with two forms. It was later discovered that in fact a gene may have more than two forms. This situation is known as multiple allelism. The members of an allelic series may exhibit any type of dominance relationship with the other members. The genes controlling the rejection of incompatible tissues in humans are an example of multiple allelism.

One gene may affect more than one character. Such genes are known as pleiotropic genes. An example is the A^Y allele mice, which affects both coat colour and survival. The identification of a major gene affecting a character does not mean that it is the only gene affecting that character: several genes may be interacting. A good example of gene interaction is the coat colour of mice, which is produced by a complex set of interacting genes that determine pigment type, pigment distribution in the hair, pigment distribution on the animal, and the presence or absence of pigment. Gene interaction often produces modified Mendelian ratios in the F_2. Some kinds of interaction have specific names, such as complementary gene action, epistasis, suppression, and duplicate gene action. Gene interaction occurs in both diploid and haploid organisms. Two other important extensions to Mendelian analysis are the concepts of penetrance and expressivity. Penetrance is the percentage of individuals of a specific genotype who express the phenotype associated with that genotype. Expressivity refers to the degree of expression, or severity, of a particular genotype at the phenotypic level.

10

EUKARYOTIC CHROMOSOME

We have previously discussed the control of gene expression in prokaryotes and bacteriophages. Compared to eukaryotes, bacteriophages and prokaryotes are relatively simple. Of fundamental importance is that, in these lower forms, the operon model of induction and repression of transcription is a unifying theme for control of gene expression. Despite nuances such as catabolite repression and attenuator control, the operon model provides a relatively clear picture of how genes are turned on and off in phages and prokaryotes. This model does not exist for eukaryotes. In attempting to elucidate models for control of gene expression in eukaryotes, we must take one very important factor into account: the complexity of the structure of the eukaryotic chromosome. In this chapter, we cover the current understanding of how these very large structures are organized.

EUKARYOTIC CELL

Eukaryotes and prokaryotes are the two superkingdoms of organisms. The following comparisons, using *E. coli* as a general model for prokaryotes, show how much more complex eukaryotes are:

1. An *E. coli* chromosome contains approximately 4.2×10^6 base pairs of DNA. The haploid human genome contains nearly one thousand times as much DNA.
2. An *E. coli* cell has very little internal structure. Eukaryotes have a number of internal organelles and an extensive lipid membrane system, including the nuclear envelope itself.
3. Eukaryotic DNA is in the form of nucleoprotein, a DNA-histone protein complex. Although a few histonelike proteins have been

found in *E. coli*, its chromosomal DNA is not complexed with protein to anywhere near the same extent.

4. An *E. coli* cell is small (0.5 to 5.0 μm in length for bacteria). Eukaryotic cells are generally larger than prokaryotes (10 to 50 μm in length for animal tissue cells).
5. The messenger RNA of *E. coli* is translated while it is being transcribed. Eukaryotic messenger RNA is modified within the nucleus before it is transported out for translation in the cytoplasm.
6. Almost no messenger RNA isolated from eukaryotic cells, including the messenger RNA of animal viruses, has been found to be polycistronic (containing many genes). Most prokaryotic messenger RNAs are polycistronic.
7. Most *E. coli* genes are parts of inducible or repressible operons; there are almost no operons in eukaryotes.
8. *E. coli* exists as a simple, single cell. Although some prokaryotes do aggregate, sporulate, and show a few other limited forms of differentiation, they are primarily one-celled organisms. And, although some eukaryotes are single-celled (e.g., yeast), the essence of eukaryotes is differentiation. In human beings, a zygote gives rise to every other cell type in the body in a relatively predictable manner.

Eukaryotic Chromosome

DNA Arrangement

Evidence that the eukaryotic chromosome is *uninemic*—that is, contains one double helix of DNA—comes from several sources.The best data are provided by radioactive-labeling studies, first done by J. Taylor and his colleagues in 1957. If a eukaryote is allowed to undergo one DNA replication in the presence of tritiated (^{3}H-) thymidine, each of the daughter chromatids would be expected to contain a double helix with one unlabeled DNA template strand and one labeled strand of newly synthesized bases. This configuration is expected on the basis of semiconservative replication, with each chromatid containing one double helix. A second round of DNA replication, in the absence of ^{3}H-thymidine, should produce chromosomes in which one chromatid would have unlabeled DNA and one would have labeled DNA. As expected, one chromatid of every pair is labeled and one is not.

In another kind of experiment, R. Kavenoff, L. Klotz, and B. Zimm demonstrated that *Drosophila* nuclei contained pieces of DNA of the size predicted from their DNA content, based on the premise

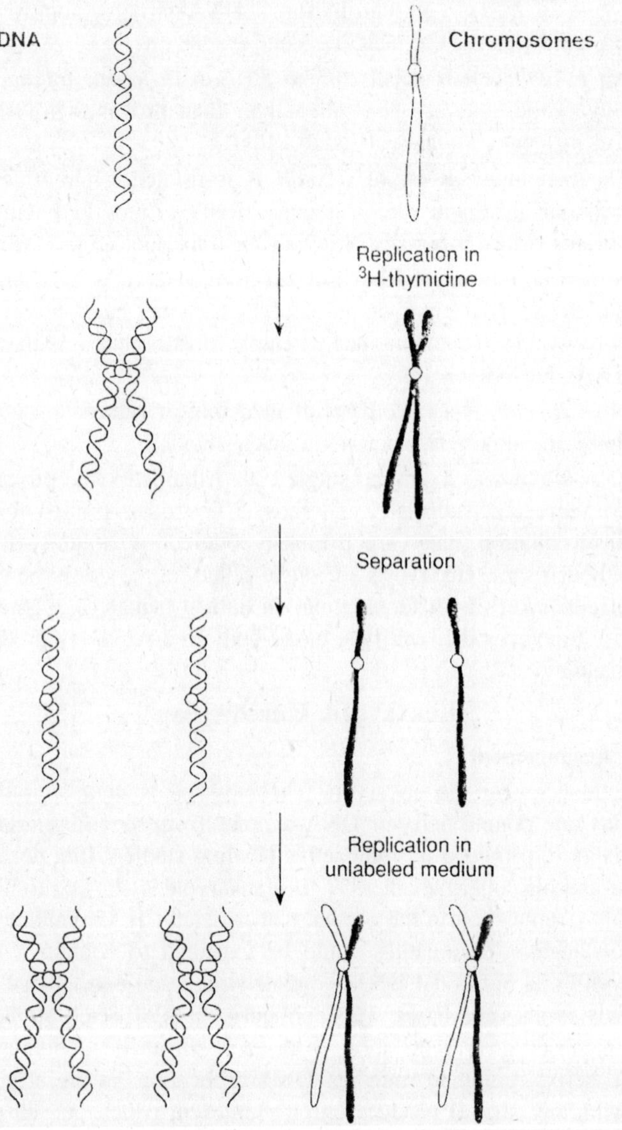

Fig. 10.1. Radioactive labeling of a uninemic eukaryotic chromosome following semiconservative replication.

that each chromosome contains one DNA molecule. They isolated the DNA and measured the size of the largest DNA molecules using the *viscoelastic* property of DNA, the rate at which stretched molecules relax. From other sources, primarily UV absorbance studies, it was estimated that the largest *Drosophila* chromosome had about 43 $\times$ 10^9 daltons of DNA. Results from the viscoelastic measurements indicated the presence of DNA molecules of between 38 and 44 $\times$ 10^9 daltons. Viscoelastic measurements of inversions, which changed the ratio of the arms but not the overall size of the chromosome, yielded similar results. However, a translocation that radically changed the size of the chromosome to 59 $\times$ 10^9 daltons resulted in an equivalent change in the viscoelastic estimates to between 52 and 64 $\times$ 10^9 daltons.

The conclusion from these studies is that the largest *Drosophila* chromosome, and by extension every eukaryotic chromosome, contains a single DNA molecule running from end to end, encompassing both arms. The viscoelastic values were corroborated by carefully isolating and measuring the lengths of long DNA molecules, an especially difficult task given DNA's propensity to break. The longest molecule that the investigators found was 1.2 cm long, equivalent to between 24 and 32 $\times$ 10^9 daltons, close to the predicted size. Thus, the evidence is in complete concordance with the simple uninemic model of eukaryotic chromosomal structure.

Nucleoprotein Composition

Nucleosome structure

Since each eukaryotic chromosome consists of a single, relatively long piece of duplex DNA, the average diploid cell contains many of these long pieces of DNA. For chromosomes to be properly distributed to each daughter cell during mitosis and meiosis, they must be condensed into structures that are more easily managed. Wrapping the DNA around "spools" of protein constitutes the first step in a series of coiling and folding processes that eventually result in the fully compacted chromosome we see at metaphase.

Interphase nuclei can be disrupted by placing them in a hypotonic liquid such as water. When this happens, chromatin material is released. When this material is observed under the electron microscope, small particles called *nucleosomes* can be seen. These are the spools that the DNA is wrapped around. They are made of *histone* proteins and associated DNA. The histones, a group of arginine- and lysine-rich basic proteins, have been well characterized. They are especially well suited to bind to the negatively charged DNA.

When chromatin is treated with micrococcal nuclease, individual nucleosomes can be isolated, indicating that the DNA between nucleosomes is accessible to digestion. The results of these studies

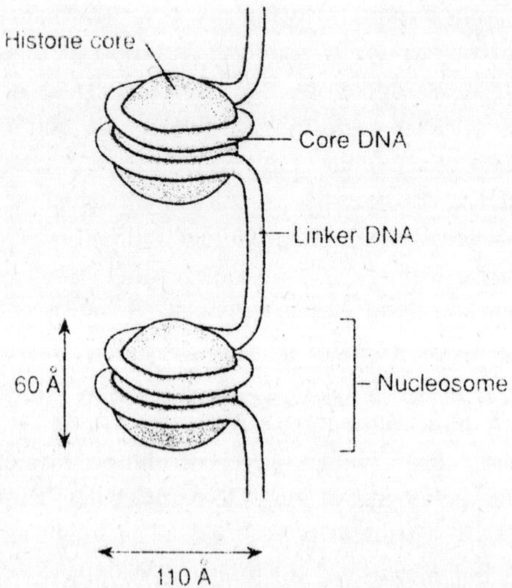

Fig. 10.2. The eukaryotic chromosome is associated with histone proteins to form nucleosomes.

indicate that a length of 168 base pairs (bp) of DNA, the core DNA, is intimately associated with the nucleosome, and another 50 to 75 base pairs, depending on species, connects the nucleosomes (linker DNA). When the quantities of the various histones were measured, there were two each of histones H2A, H2B, H3, and H4 per nucleosome and only one molecule of histone H1. Reconstitution and degradation studies have indicated that histone H1 is not a necessary component in the formation of nucleosomes. We believe that histone H1 is associated with the linker DNA as it enters and emerges from the nucleosome, although its exact position is not known with certainty. Histone H1 may be more off center and internally located than illustrated. The term *chromatosome* has been suggested for the core nucleosome plus the H1 protein, a unit that includes approximately 168 base pairs of DNA. Nucleosomes, then, are a first-order packaging of DNA; they reduce its length and undoubtedly make the coiling and contraction required during mitosis and meiosis more efficient.

When DNA is replicated, twice as many nucleosomes are needed since one double helix becomes two. Recent studies indicate that a parental nucleosome is partly disassembled during DNA replication and reassembled on one or the other daughter strand, apparently randomly. The other DNA strand has a new nucleosome constructed of histones from the cellular pool with the help of proteins called *chromatin assembly factors*; at least three of these factors are known.

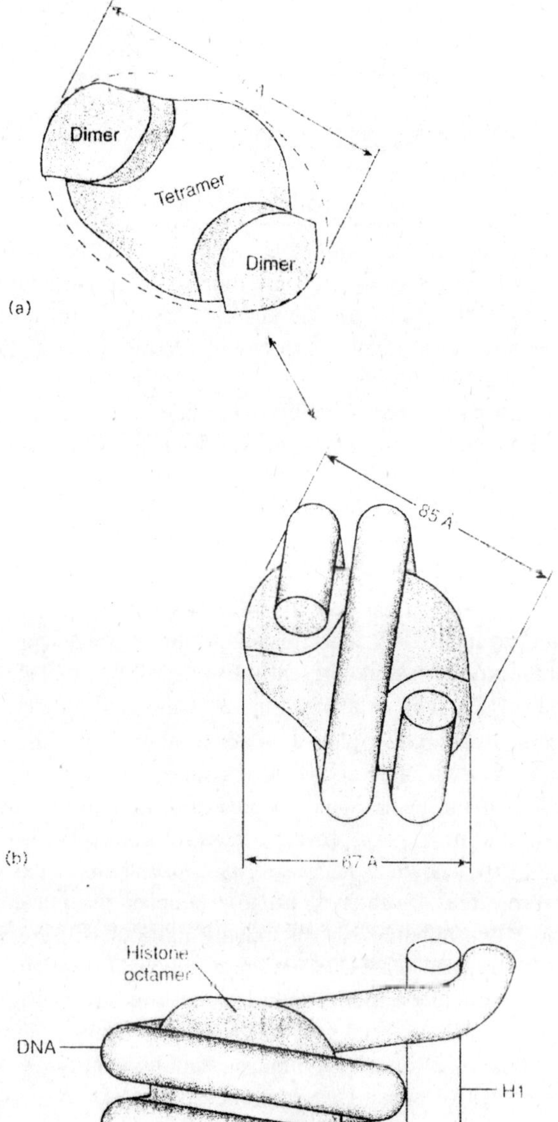

Fig. 10.3. Nucleosome structure.

Table 10.1. The constituency of calf thymus chromatin

Constituent	*Relative Weight*
DNA	100
Histone proteins	114
Nonhistone proteins	33
RNA	1

For example, in fruit flies, a protein complex called the *replication coupling assembly factor* assembles new nucleosomes. In addition, a protein complex called *condensin* is needed for the condensation of interphasechromosomes to mitotic chromosomes. This complex includes two *SMC proteins* (for structural maintenance of chromosomes) and two non-SMC proteins. SMC proteins also aid other chromosomal activities, such as mitotic segregation, sister-chromatid adhesion, dosage compensation, and recombination. Thus, a diverse array of proteins is involved in creating nucleosomes and chromatosomes, condensing interphase chromosomes, and performing numerous other activities of chromosomes.

Dosage compensation has recently been associated with a change in nucleosome structure. The inactivated X chromosome appears to have a different type of histone present. Histone H2A is replaced by a variant called mH2A. The details of this mechanism are under study.

Nucleosomes apparently play a major role in controlling gene expression; DNA with nucleosomes has a much lower transcription rate than DNA without nucleosomes. It makes sense that the positions of nucleosomes can provide or prevent access to promoters. There are regions of the DNA, known as *nuclease-hypersensitive sites*, that appear to be nucleosome free. These sites, usually mutiples of a nucleosomal region of about two hundred base pairs, are particularly sensitive to digestion by different nucleases. When these regions are isolated, they usually have sequences that control functions in replication, transcription, or other activities of DNA. For example, numerous promoter regions in *Drosophila*, mouse, and human DNA are in nuclease-hypersensitive sites. Hence, some specific DNA sequences are kept free of nucleosomes, and these sequences appear to be recognized by various enzymes such as RNA polymerase. In many other cases, however, nucleosomes do appear to cover promoters and repress transcription. For transcription to occur in these cases, some form of *chromatin remodeling* must take place.

Two general classes of proteins are involved in chromatin remodeling. First are proteins that acetylate the N-terminal tails of the histones, a process that may cause the nucleosomes to bind the DNA less tightly and thus make it available for attachment of transcription factors. These enzymes are called *histone acetyl transferases* (HATs). Deacetylating enzymes have the reverse effect: They act to repress transcription. Second, a class of ATP-dependent proteins such as the SWI/SNF complex in yeast also affect chromatin remodeling. (Some workers called the proteins SWI because they were involved in mating type switching, and others called them SNF for sucrose nonfermenting.) The SWI/SNF complex is a group of eleven proteins involved in transcription activation in many genes, presumably allowing transcription factors to access promoters by remodeling chromatin. These proteins are able to reposition a nucleosome on DNA by sliding the nucleosome down the DNA.

We thus conclude that although nucleosomes serve as a general, first-order packing mechanism in eukaryotic DNA, they can be positioned precisely and can attenuate transcription. It is interesting to note that once transcription begins, RNA polymerase apparently moves along nucleosomed DNA by translocation of the histones by 75 to 80 base pairs without disrupting the nucleosome itself. This seems to be accomplished by the RNA polymerase moving the DNA and then re-forming the nucleosome in its wake.

Higher-order structure of chromatin

Since the nucleosome has a width of only 110 Å, and metaphase chromosomes appear to be constructed of a fiber having a diameter of about 2,400 Å, several additional levels of chromatin compaction lead to the metaphase chromosome. Various experiments, which change the ionic strength the chromatin is subjected to, indicate that the 110 Å DNA spontaneously forms a 300 Å, solenoidlike fiber with increased ionic strength. It seems that this fiber results from the coiling of the nucleosomal DNA. This 300 Å fiber is not, however, the final form of the DNA. We can account for the contraction of the 300 Å fiber to the 2,400 Å fiber found in metaphase chromosomes by the formation of a second solenoidlike structure from the winding of the 300 Å fiber.

If the histones are removed from a chromosome, the DNA billows out, leaving a proteinaceous structure termed a *scaffold*. This scaffold structure is formed from *nonhistone proteins*; two of them predominate, namely SC1 and SC2. SC1 has been identified as topoisomerase II. It

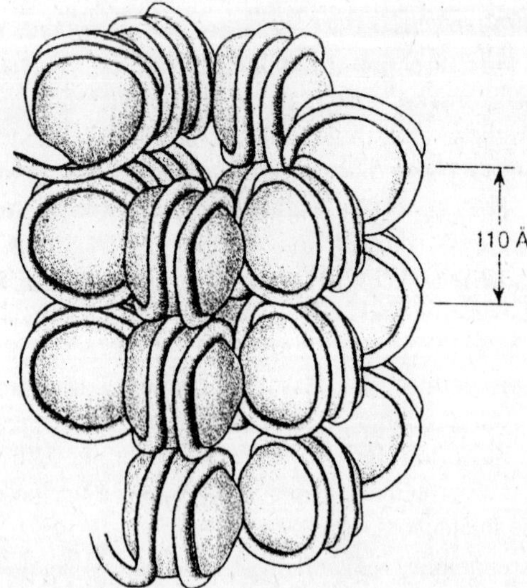

Fig. 10.4. Solenoid model for the formation of the 300-Å chromatin fiber.

would not be unreasonable to expect several hundred different proteins, many in minute quantities, to be associated with the chromosome and involved in replication, repair, and transcription.

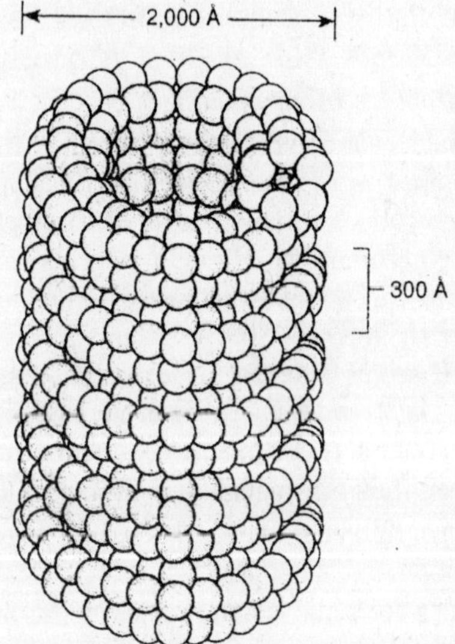

Fig. 10.5. The 2,000-Å fiber of the eukaryotic chromosome is a solenoidlike structure formed by the coiling of the 300-Å fiber, which itself is a solenoid.

Polyteny, puffs, and balbiani rings

Drosophila's salivary glands, as well as some other tissues of *Drosophila* and other diptera, contain giant banded chromosomes that result from the replication of the chromosomes and the synapsis of homologues without cell division (endomitosis). These chromosomes

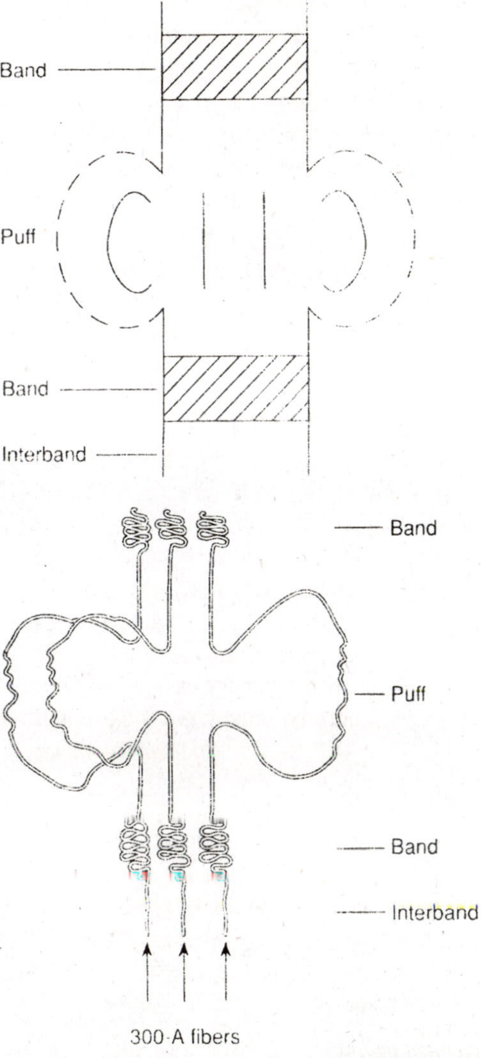

Fig. 10.6. Polytene chromosome with bands and a puff. Three of the approximately one thousand synapsed chromatids are shown diagrammatically on the below.

consist of more than one thousand copies of the same chromatid and appear as alternating dark bands and lighter interband regions. The dark bands are referred to as chromomeres. Also seen are diffuse areas called *chromosome puffs*. Chromosome puffs are also referred to as *Balbiani rings*. These rings were originally defined as puffs in the midge, *Chironomus*, whose polytene chromosomes were discovered by E. G. Balbiani in 1881. Currently, the term applies to all puffs, or at least the larger puffs, in all species with polytene chromosomes.

The structure of the polytene chromosome can be explained by the diagram. Dark bands (chromomeres) are due to tight coiling of the 300 Å fiber; light interband regions are due to looser coiling. The figure also shows how chromosome puffs would come about as fibers unfold in regions of active transcription.

Staining with reagents specific for RNA, such as toluidine blue, or autoradiography with tritiated (^{3}H) uridine, have been used to demonstrate that active transcription is going on in the puffs but not in neighbouring regions of the polytene chromosomes. The messenger RNA isolated from cells with puffs has also been shown to hybridize only to the puffed regions of the chromosomes. Thus, these regions of the DNA are complementary to the messenger RNA and represent areas of active transcription. Modern recombinant DNA techniques have also shown that many puffs probably represent the transcription of only one gene, although there are exceptions.

Puffs generally fall into four categories. *Stage-specific puffs* appear during a certain stage of development, such as molting. *Tissue-specific puffs* are active in one tissue but not another. (In dipteran larvae, tissues other than the salivary glands, such as the midgut and Malpighian tubules, have polytene chromosomes.) *Constitutive puffs* are active almost all the time in a specific tissue. And *environmentally induced puffs* appear after some environmental change, such as heat shock. In *Drosophila*, about 80% of the puffs are stage specific; in *Chironomus*, only about 20% are. For example, at the time of molt in insects, the hormone ecdysone is secreted by the prothoracic gland. At the same time, many puff patterns change. Similar changes in puff patterns can be induced by the injection of ecdysone. Hence, molting, a stage-specific developmental sequence, is related to a sequential transcription sequence in the chromosomes.

Lampbrush chromosomes

Lampbrush chromosomes, which occur in amphibian oocytes, are so named because their looped-out configuration has the appearance of a brush for cleaning lamps, now a relatively uncommon household item. The loops of the lampbrush chromosomes are covered by an RNA matrix and are the sites of active transcription. Presumably, the loops are unwindings of the single chromosome, similar to the unwindings in the polytene chromosome. Thus, under certain circumstances, such as in polytene chromosomal puffs and in lampbrush chromosomes, active transcription can be seen in the light microscope. Since only certain bands puff at any one moment in polytene chromosomes, and since the loops of lampbrush chromosomes are of various sizes (with some regions not looped at all), we have evidence of specific transcription. However, we have no indication, so far, of the nature of the control of that transcription.

Chromosomal Banding

Several chromosomal staining techniques reveal consistent banding patterns. By means of these patterns, all of the human chromosomes can be differentiated. Of possibly greater importance is the fact that these staining techniques have provided some insight into the structure of the chromosome. The techniques for staining the C, G, and R chromosomal bands will serve as an illustration.

G-bands are obtained with *Giemsa stain*, a complex of stains specific for the phosphate groups of DNA. Treatment of fixed chromatin with trypsin or hot salts brings out the G-bands. Giemsa stain enhances banding that is already visible in mitotic chromosomes. The banding pattern is caused by the arrangement of chromomeres. Under careful observation, the major G-bands prove to consist of many smaller chromomeres. This banding appearance has led D.

C-bands are Giemsa-stained bands after the chromosomes are treated with NaOH. The *C* is for "centromere," because these bands represent constitutive heterochromatin surrounding the centromeres. The DNA is also usually satellite rich. *Satellite DNA* differs in *buoyant density* from the major portion of cellular DNA. When eukaryotic DNA is isolated and centrifuged in CsCl, forming a density gradient, the majority of the DNA forms one band in the gradient at a single buoyant density. The buoyancy is determined by the G-C content of the DNA. However, smaller secondary bands are also usually present, indicating regions of DNA having sequences different from the majority of the cell's DNA. DNA isolated this way is referred to as *satellite DNA* because of the secondary, or satellite, bands formed in the density gradient. As we will see, this DNA is found primarily around centromeres and consists of numerous repetitions of a short sequence.

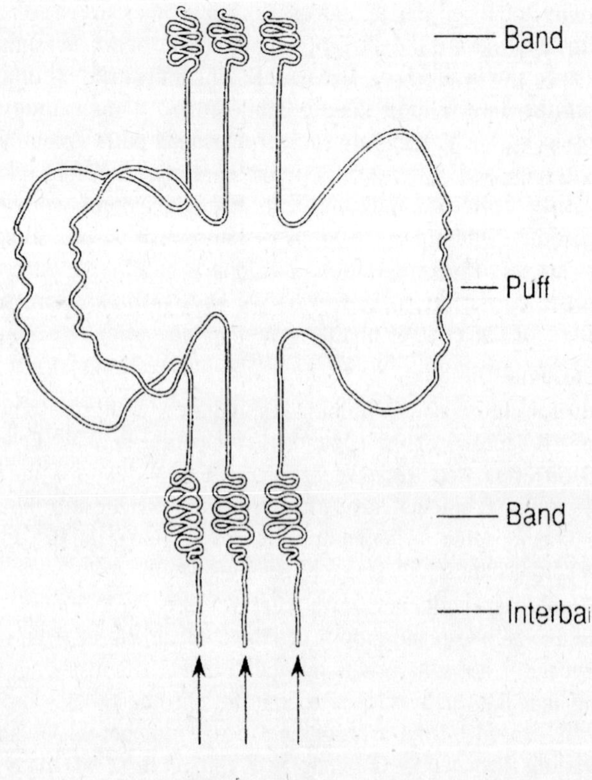

Fig. 10.7. Model of eukaryotic (mammalian) chromosomal banding.

R-bands are visible with a technique that stains the regions between G-bands. The chromosomes are fixed, stained with Giemsa, and then viewed with a phase contrast microscope. Since the dark-light pattern is the opposite of the G-band pattern, these bands are called *reverse bands*.

From the information gleaned from these staining techniques, D. Comings distinguished between three basic chromatin types: euchromatin, constitutive heterochromatin, and intercalary heterochromatin. Presumably, the only chromatin involved in transcription is *euchromatin*. *Constitutive heterochromatin* surrounds the centromere and is rich in satellite DNA. *Intercalary heterochromatin* is found throughout the chromosome. Thus, it becomes apparent that the eukaryotic chromosome is a relatively complex structure.

Centromeres and Telomeres

Centromeres

Two regions of the eukaryotic chromosome have specific functions—the centromere and the telomeres. The centromere is involved in chromosomal movement during mitosis and meiosis, whereas the telomeres terminate the chromosomes. As we pointed out, the terms *centromere* and *kinetochore*, while occasionally used interchangeably, are distinct. The kinetochore is the interface between the visible constriction in the chromosome (the centromere) and the microtubules of the spindle. The kinetochore of higher organisms (e.g., mammals) contains proteins and some RNA. Microscopically, it is a trilaminar structure, attached to chromatin at the inner layer and to microtubules at the outer layer.

Most of our knowledge of the genetics of centromeres has come from work in yeast (*Saccharomyces cerevisiae*). Cells did not maintain most artificially created yeast plasmids because they were lost during mitosis. However, plasmids were isolated that did replicate normally during cell division. Presumably, they contained centromeres, allowing them to replicate and move in synchrony with the host's chromosomes. Further genetic engineering made it possible to isolate smaller and smaller regions that could serve as centromeres. After sequencing the centromeres of fifteen of the sixteen yeast chromosomes, it was possible to conclude that the centromere from yeast is about 250 base pairs long with three consensus regions; we are defining a centromere as a sequence of DNA called the *CEN* locus or CEN region. Recent data indicate that this region may contain a single, modified nucleosome associated with region II. The 250 base-pair length of the CEN regions of yeast chromosomes is about 200 Å, the same as the diameter of a microtubule, indicating that only one micro-tubule attaches to each centromere during mitosis or meiosis in a yeast cell. This region is called a *point centromere*. Higher eukaryotes have larger centromeric regions that attach more microtubules. These regions are referred to as *regional centromeres*. Regional centromeres range from nineteen to one hundred kilobases (kb; 19,000–100,000 bases) with unique and satellite (repeated sequence) DNA that is heterochromatic and may include expressed genes. We know much less about regional centromeres than we do about point centromeres.

Telomeres

Since eukaryotic chromosomes are linear, each has two ends, referred to as *telomeres*, that not only mark the termination of the linear chromosome but also have several specific functions. Telomeres must prevent the chromosomal ends from acting in a "sticky" fashion, the way that broken chromosomal ends act. In other words, chromosomal ends must not elicit a DNA repair response. Telomeres must also prevent the ends of chromosomes from being degraded by exonucleases and must allow chromosomal ends to be properly replicated.

Most telomeres isolated so far are repetitions of sequences of five to eight bases. In human beings, the telomeric sequence is TTAGGG, repeated 300 to 5,000 times at the end of each chromosome. The human telomere was discovered by R. Moyzis and his colleagues when they probed the highly repetitive segment of human DNA. (Highly repetitive DNA, as its name implies, consists of numerous copies of a single sequence and usually comprises the satellite components of the cell's DNA; see next section.) When a probe for this sequence was applied to human chromosomes, the sequence was found at the tip of each chromosome in roughly the same quantity.This is a highly conserved sequence, found in all vertebrates studied as well as in unicellular trypanosomes. Similar sequences are found in various other eukaryotes; the first sequence was isolated by E. Blackburn and J. Gall in 1978.

When a linear DNA molecule is replicated, the 3' $\rightarrow$ 5' strand can be replicated to the end. The 5' $\rightarrow$ 3' strand, however, is replicated with RNA primers that are then degraded, leaving a short gap on the progeny strand. It is always the G-rich strand of telomeric DNA that ends up single-stranded, forming a 3' overhang of twelve to sixteen nucleotides. Thus, the normal replication process of a linear DNA molecule leaves an incomplete terminus. Hence, scientists suspected that there would be a unique mechanism for the replication of telomeres.

Telomeric sequences appear to be added de novo without, DNA template assistance by an enzyme called *telomerase*, discovered by E. Blackburn and her colleagues. This was seen when telomeres from another species were engineered into yeast cells. After a cell cycle, the yeast telomeric sequence had been added on at the ends of the foreign chromosome, the result, presumably, of the telomerase enzyme.

When Blackburn and her colleagues isolated telomerase, they discovered that a segment of RNA, about 160 base pairs, is an integral part of the enzyme. That RNA has a region that is complementary to the G-rich repeat of the telomeric DNA sequence of the species. After careful experimentation, including modifying the gene for the

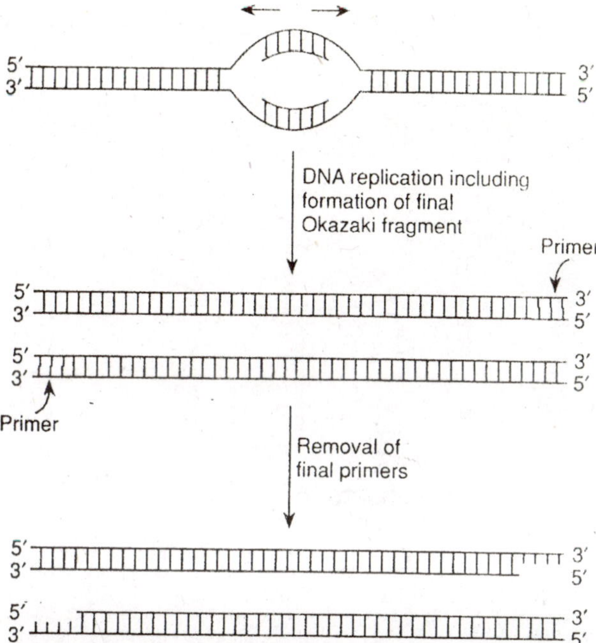

Fig. 10.8. Removal of final primers after the replication of linear DNA creates single-stranded ends.

telomerase RNA, Blackburn and her colleagues concluded that telomerase uses its RNA as a template for adding telomeric repeats to the ends of chromosomes. Telomerase is thus a reverse transcriptase, using RNA nucleotides as a template to polymerize DNA nucleotides.

Blackburn and her colleagues proposed that the first step in telomere extension is hybridization of the 3' end of the telomere with the RNA component of telomerase. Then, with the telomerase RNA as a template, the 3' end of the telomere is extended. Finally, a translocation step takes place that displaces the telomere in respect to the RNA, returning to the configuration at the beginning of the process. The single-stranded C-rich strand is then synthesized with DNA polymerase and DNA ligase.

Once telomeres have been added to the ends of eukaryotic chromosomes, different organisms use any of three different methods known to protect the ends of the chromosomes. First, the guanine-rich DNA can form complex structures. Biochemists have discovered that four guanines can form a planar *G-tetraplex*, with the four bases hydrogen bonded to each other. Several structures have been hypothesized

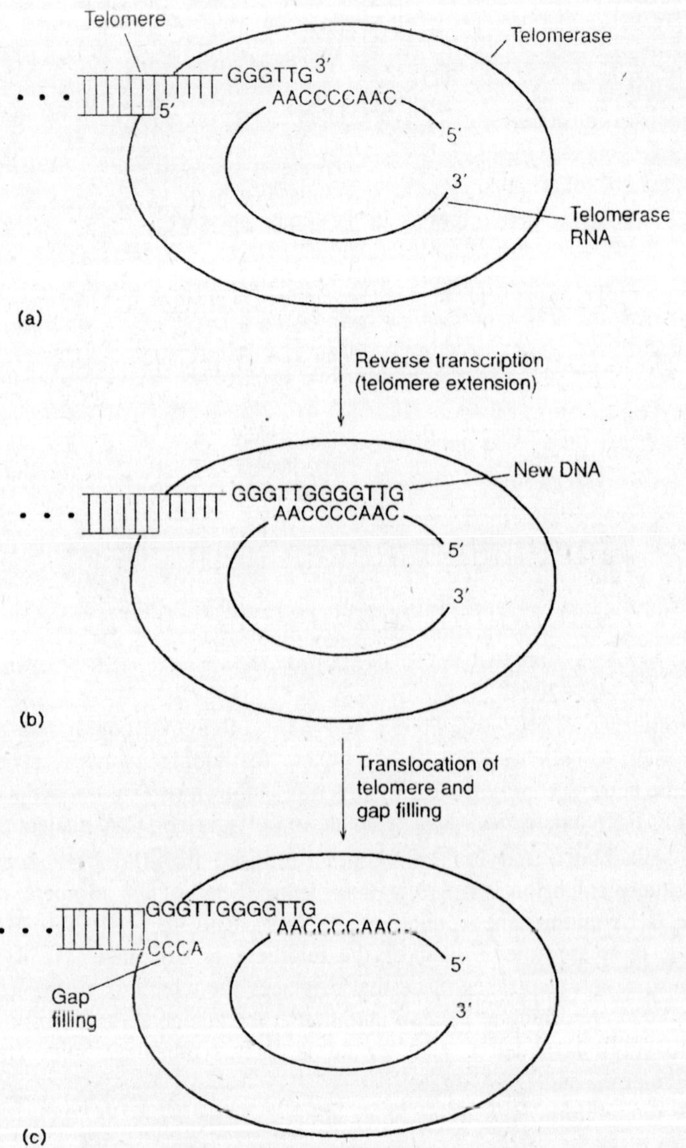

Fig. 10.9. Telomerase extends telomeres using telomerase RNA as a template Gap filling by DNA polymerase I and ligase complete the double helix.

to explain the novel ends of these chromosomes. Second, proteins have been discovered that bind to the 3' ends of telomeres. In the ciliate *Oxytricha nova*, a protein called the telomere end-binding protein (TEBP) attaches to the 3' ends of telomeres and protects them. Finally, a novel structure called the *t-loop* has been discovered at the end of mammalian telomeres. This loop forms at the ends of chromosomes under the direction of a protein called TRF2 (telomere repeat-binding factor), which causes the 3' end of the chromosome to loop around and interdigitate into the double helix, forming the loop.

How do cells keep track of the number of their telomeric repeats? Proteins have been isolated that bind to telomeres (Rap1 in *Saccharomyces cerevisiae*, TRF1 in human beings). By mutating these proteins or the telomeric sequences, scientists have changed the equilibrium number of telomeric repeats.This led to the current model that the cell counts the number of these proteins bound to the telomeres, not the number of telomeres directly, to know whether telomeres should be added. This is a very active area of research.

In yeast, protozoa, and other single-celled organisms, telomerase is active, keeping the ends of the chromo somes at the appropriate lengths. These cells can divide potentially forever. However, in most cells of higher organisms, telomerase is not active, and the ends of the chromosomes get shorter with each cell division. At a certain telomeric length, the cells no longer divide. However, if telomerase becomes active, and the ends of the chromosomes lengthen, a signal is conveyed to keep cells dividing, which can lead to cancerous growth. In fact, human telomerase was isolated from an immortal cell line (HeLa) derived from cervical cancer cells. Thus, attention is now turning to the possible clinical application of this knowledge: If telomerase can be deactivated in tumor cells, the cells may stop dividing or die, thereby eliminating the cancer. Further, studying normal telomere shortening, which appears to act as a biological clock, may help us understand the aging process and senescence.

C-Value Paradox

Why do eukaryotes have so much DNA, and why is there huge variation in the DNA content between species of comparable complexity? These questions define the *C-value paradox*, in which *C* refers to the quantity of DNA in a cell. For an example of the paradox, although human beings have 3.3 billion base pairs in the haploid genome, an amoeba has more than 200 billion base pairs. And although an average bony fish has over 300 billion base pairs of DNA in its haploid genome, the Japanese puffer fish has less than half a billion base pairs. If the basic bony fish pattern can be created with less than half a billion base pairs, why does the average bony fish have over 600 times that much DNA? What is this excess DNA doing? To explain the C-value paradox, researchers examined the repetitiveness of DNA, and more recently, probed and sequenced DNA to understand its properties.

DNA-DNA Hybridization

R. Britten and his colleagues, using the technique of *DNA-DNA hybridization*, first systematically analyzed the repetitiveness of the DNA within eukaryotes. When DNA is heated, it denatures or unwinds into single strands; when it cools, it renatures. The rate of renaturation depends on the DNA sequences. If the sample contains DNA with repeated sequences, it will hybridize faster than DNA that does not have repeated sequences. From these studies, Britten and his colleagues found that eukaryotic chromosomes contain regions of unique, moderately repetitive, and highly repetitive DNA. *Unique DNA* is, as its name implies, DNA with unrepeated sequences. *Repetitive DNA* is DNA whose sequences are repeated in the genome.

Satellite DNA, found around centromeres, is highly repetitive DNA with a unique repeat length of about two hundred base pairs. Given the quantity of satellite DNA per cell, there must be more than one million repetitions of this two-hundrednucleotide sequence in higher eukaryotes. At the other end of the spectrum is unique DNA, which makes up most of the transcribed genes of an organism. The rest of the DNA is repetitive DNA in a few to several hundred thousand copies. This repetitive DNA comprises at least three categories. One is "junk" DNA, DNA that is not useful to the organism, made up of untranscribed and parasitic sequences (selfish DNA). Another category is transcribed genes in many copies that have diverged from each other, such as antibody, collagen, and globin genes. We use the term *gene family* to refer to genes that have arisen by duplication, with or without divergence, from an ancestral gene. And finally, transcribed genes in many copies that are virtually identical, such as ribosomal RNA and histone genes, make up a third category of repetitive DNA.

Junk DNA

We saw that transposons in prokaryotes are generally viewed as selfish or parasitic: They serve no purpose to the cell. The transposons replicate on their own, increasing in number. Eukaryotic transposons are mostly *retrotransposons*, transposable elements that move by way of an RNA intermediate. That is, the retrotransposon is transcribed into RNA and then, by reverse transcription, converted to a cDNA that is then inserted into the genome. These elements can make up

50% of the eukaryotic genome, existing in hundreds of thousands of copies. They generally fall into two categories: LINES and SINES. *Long interspersed elements* (LINES), are up to seven thousand base pairs each and contain genes for reverse transcription, RNA binding, and endonuclease activity. They thus have the ability to jump by way of an RNA intermediate. Human DNA is believed to be composed of about 15% LINES.

Short interspersed elements (SINES) are generally derivatives of transfer RNA genes and do not have the ability to retrotranspose on their own. That is, in the past, their transcripts were modified, converted to cDNA by reverse transcription, and then reinserted into the host's genome. They rely on the reverse transcriptase provided by the genes of LINES or retroviruses. One group of SINEs not derived from transfer RNA is derived from the RNA of the signal recognition particle; members of this group occur in human beings in about five-hundred thousand copies of a three-hundred-basepair sequence. Because these sequences are cleaved by the restriction endonuclease *Alu*I, they are called the *Alu family*. The human genome is also permeated by remnants of at least a dozen distinct families of ancient retroviruses scattered throughout our chromosomes.

At this point, we can see some potential explanations for the C-value paradox. Much eukaryotic DNA is junk, apparently doing no harm. In some cases, 97% of the host genome is composed of junk DNA. Recent work seems to indicate that gross differences in DNA content between higher organisms may be due to the differing abilities of different species to rid themselves of this parasitic DNA. If it builds up without being removed, the DNA content of the species can soar. Thus, the wide differences in DNA content among higher eukaryotes mentioned at the beginning of this section have little to do with the complexity of the organism, but rather with the ability of the organism to remove junk DNA as it forms.

Expressed genes in many copies

Several types of genes create a product that is needed in such large quantity that one copy of the gene could not fulfill the cell's needs. We are familiar with the nucleolus, the site of the ribosomal RNA genes. Human beings have about two hundred copies of the major ribosomal RNA gene and about two thousand copies of the 5S ribosomal RNA gene. Fruit flies have about two hundred and one hundred copies, respectively, of the two genes. In some cases, the normal number of multiple copies of a gene is still not enough. The cell must then resort to *gene amplification*, a process whereby the cell increases the number of copies of the gene. For example, during oogenesis, ribosomal RNA genes (rDNA) are often amplified. In *Xenopus*, rDNA is amplified about one thousand times, which allows an oocyte to accumulate about 10^{12} ribosomes. The amplified DNA is in the form of small, circular, extrachromosomal molecules of DNA. Several models have been proposed as to how cells actually amplify their DNA. One model relies on unequal crossing over (as in *Bar* eye in *Drosophila*), whereas another model is based on unscheduled extra DNA replication in a region, followed by recombinational events that generate linear and circular forms of the excess DNA. It is not presently clear which model is correct.

In addition to ribosomal RNA genes, other genes are repeated, ensuring adequate gene products. The number and location of repeated genes are usually discovered by hybridization studies using probes, similar to the way that telomeric DNA was shown to be at the tips of the chromosomes. Repeated genes include the genes for transfer RNAs and histones. The average transfer RNA is repeated about a dozen times in *Drosophila*. Human beings have over thirteen hundred copies of transfer RNA genes in the haploid genome. In many species, the five histone genes form a repeated cluster, although each gene is transcribed independently, while prokaryotic operons are transcribed as a unit. The arrangement of histone genes may be more complex in higher forms. There are indications that in mammals, histone genes may lie in small groups or even as individual genes.

Several types of genes occur in similar but not identical forms—that is, an original gene was duplicated but, unlike histone or ribosomal RNA genes, the copies diverged in function. These gene families include globin genes, immunoglobulin genes, chorion protein (insect eggshell) genes, and *Drosophila* heat shock genes.

Globin gene family

Globins are oxygen-transporting and storage molecules found in animals, some plants, and microorganisms. In higher vertebrates, there are two types of globins: myoglobin, which stores oxygen in muscles, and hemoglobin, found in red blood cells. Myoglobins function as single molecules, whereas hemoglobins occur as tetramers, two each of two protein chains. Evolution in the globin gene family can be traced by comparative studies of globins in different species as well as molecular studies of globins within a species. Studying hemoglobins has provided a great deal of information on gene expression and evolution. We turn our attention to the globin gene family in human beings. During human development, four major hemoglobins appear: embryonic hemoglobin, Hb F, Hb A, and Hb A_2. Structurally, the ζ (Greek, zeta) subunit (a component of embryonic hemoglobin) is α-like, whereas the rest are β-like. Fetal hemoglobin has a higher affinity for oxygen than does adult hemoglobin, thus allowing fetuses to draw oxygen from their mother's blood. From a comparative study of the DNA sequences, the evolution of the various hemoglobin genes has been inferred.

The α genes are located in a cluster on chromosome 16; the β genes are located in a cluster on chromosome 11. These two clusters provide a clear case history of gene duplication, presumably by unequal crossing over, followed by divergence. Having a second or third copy of a gene allows one of the duplicates to diverge (and perhaps to become nonfunctional in the process), whereas the original still performs the required function.

Many diseases of genetic interest involve the hemoglobins. In fact, hemoglobinopathies, including sickle-cell anemia and the thalassemias, are the most common genetic disorders in the world population. The best-known mutation of a hemoglobin gene itself is the one that causes sickle-cell anemia, a mutation of the sixth amino acid of the β chain. In the homozygous state, the disease is usually fatal. However, heterozygotes show an increased resistance to malaria. One of the ramifications is that the sickle-cell allele is maintained at relatively high frequencies in malarial regions.

The *thalassemias* are a group of diseases that affect the regulation of the α and β hemoglobin genes. (*Thalassemia* comes from the Greek for "sea blood," because the disease is best known in individuals living around the Mediterranean Sea.) In α and β thalassemias, the α or β subunit, respectively, is present in very low quantities or entirely absent. Many of the genetic defects are deletions, possibly due to unequal crossing over within the globin gene complexes. T. Maniatis showed that β thalassemia is caused by a mutation in the β-globin gene that disrupts RNA splicing. The body compensates by forming γ_4 or β_4 hemoglobin in a thalassemias, or $\alpha_2\gamma_2$ or $\alpha_2\delta_2$ in β thalassemias. These are relatively unsuitable or inefficient responses; the diseases range from very mild to very severe and frequently fatal. More information is needed regarding the control of hemoglobin production in the thalassemias.

11

CHROMOSOME THEORY OF INHERITANCE

Dolores Becker was 37 years old and decided that if she was ever going to have a child, she should not wait any longer. She soon became pregnant and happily anticipated the arrival of her child. However, the birth of her daughter touched off legal battles that still reverberate through the staid halls of New York's court houses. Mrs. Becker's daughter was born with a genetic defect known as *Down syndrome*. This condition is caused by an extra *chromosome*, a carrier of the genetic information we inherit from our parents. Mrs. Becker's physician did not warn her that, at her age, she ran a substantial risk (about 1 in 250) of giving birth to a child with Down syndrome. He could have analyzed the fetus's chromosomes to see if it had the correct number, but he did not. The Beckers sued the physician for the damages they sustained as a result of his failure to inform them of the risks and to offer diagnostic tests. In this "wrongful birth" suit, the Beckers claimed that they were denied the opportunity to make an informed decision to continue the pregnancy or terminate it. The child, through her parents, also sued the physician. It was her contention that she should never have been allowed to be born with the severe mental and physical abnormalities that characterize Down syndrome. This is called a "wrongful life" suit.

On the first go-around, the court dismissed the case, stating that it did not recognize a cause of action for "wrongful birth," as the parents claimed, or for "wrong full life," as the child claimed. The Beckers appealed the ruling to a higher court, and it was reversed.

The physician then appealed to New York's highest court of appeals. This court supported the concept of "wrongful birth," but rejected the notion of "wrongful life." So to this day in the state of New York, parents can sue a physician for a "wrongful birth," but a child cannot sue for being born in a defective condition. In the court's words:

Whether it is better never to have been born at all than to have been born with even gross deficiencies is a mystery more properly left to the philosophers and theologians.

Though New York does not recognize "wrong full life" suits, other states do. In California, a young girl was born with Tay-Sachs disease because the laboratory that tested the parents said that they were not carriers. The laboratory was negligent. The parents successfully sued the laboratory on behalf of their child (the suit was based on emotional stress and deprivation of 72.6 years of life).

The foundation on which the Becker story rests is the chromosome, a linear array of genes found in cell nuclei. Down syndrome is caused by a mistake in cell division that leads to an extra copy of chromosome number 21.

When Mendel formulated the laws governing inheritance, he knew nothing about chromosomes. Nobody did. People knew about cells, and toward the latter part of the nineteenth century they knew that the nucleus was the key to inheritance. But what entities inside the cell's nucleus provided a physical basis for the laws of inheritance? Soon after biologists discovered the pioneering experiments carried out by Mendel and the principles of inheritance that emerged from those experiments, chromosomes were identified a the cellular entities responsible for inheritance.

Cell Division and the Transmission of Chromosomes

Identifying Chromosomes as the Carriers of Heredity

Biologists' knowledge of internal cell structure was greatly enhanced by improvements in the microscope and by the development of staining techniques during the 1840s. Because various parts of the cell were found to react with specific stains, or dyes, it became possible to see structures previously not visible. Biologists soon discovered that the cell nucleus contains a threadlike substance that stains darkly with certain stains. This substance they called *chromatin* (from the Greek *chromos*, colour). Over the next few decades, many observers noticed that just before a cell divides, the *chromatin* condenses and forms dark-staining bodies called *chromosomes* ("colour bodies").

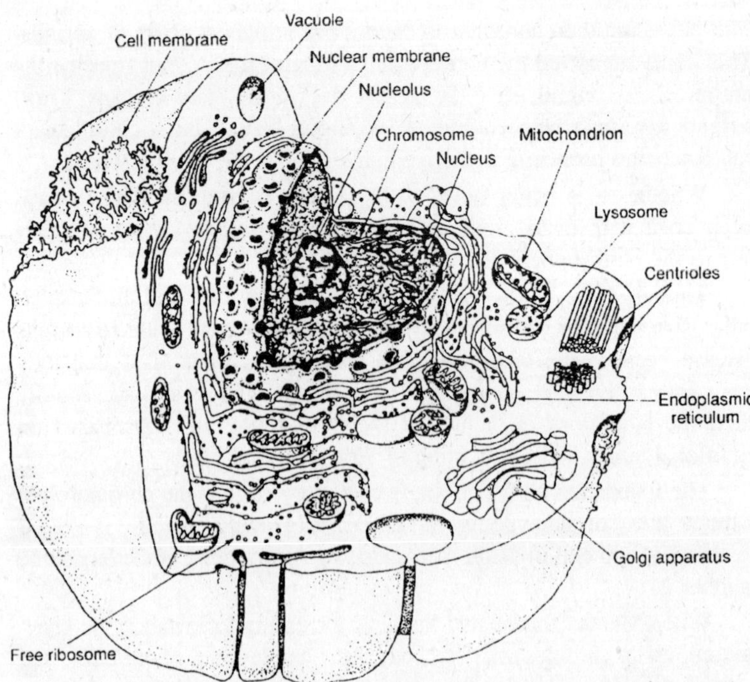

Fig. 11.1. A centralized animal cell showing some of the structures commonly found.

Mitosis: Duplication of nuclear material

By the mid-1870s cytologists realized that chromatin did not actually disappear before cell division, but condensed to form the

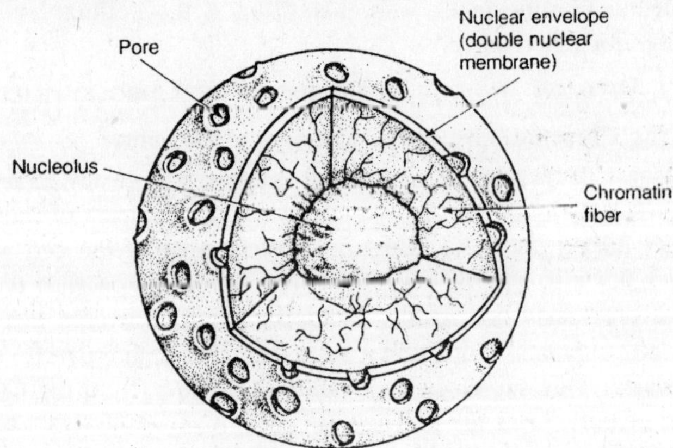

Fig. 11.2. A cell nucleus showing a nucleolus, the nuclear pores in the membrane, and the uncoiled chromatin that condenses into chromosomes during cell division.

chromosomes. They had also observed that in the early stages of cell division (prophase and metaphase) each chromosome is a double structure consisting of two strands, or *chromatids*, joined at a constricted region called the *centromere* (also called a *kinetochore*). During cell division the centromere of every chromosome splits, and the two chromatids, now called chromosomes, migrate to opposite pole of the cell. This division of the nuclear material, which occurs whenever a *somatic cell*, or non-sex cell, divides to produce other somatic cells, was later named *mitosis* (Greek *mitos*, thread) because of the threadlike appearance of the chromosomes under the microscope.

At the end of mitosis, each *daughter cell* contains undoubled chromosomes, equal in number to the double chromosomes of the parental cell. The new chromosomes then uncoil, but when the daughter cells are ready to divide, their chromosomes condense and reappear as double structures. As the nineteenth-century cytologists suspected, chromosomes are replicated between mitotic divisions. This means not

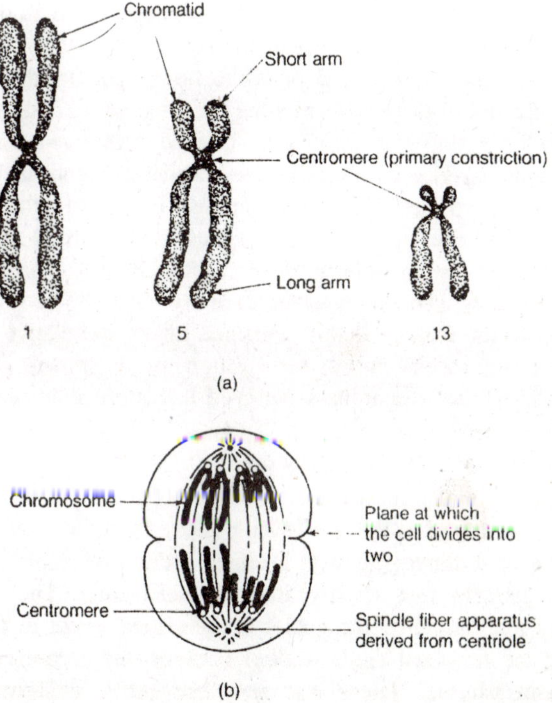

Fig. 11.3. (a) Human chromosomes, 1, 5, and 13. (b) In cell division, the centromere divides and the chromosomes move to opposite poles of the spindle.

only that the amount of genetic material is doubled but also that every gene is replicated exactly. Each of a chromosome's chromatids is identical to the other.

We will examine mitosis in more detail later in the chapter, but its essential features are simple. As a cell divides, its replicated chromosomes move apart so that each new cell has a representative of that particular chromosome. Daughter cells are genetically identical to each other and to the parental cell.

The discovery of mitosis strongly supported the idea that nuclear material is continuous from one cell generation to the next and that it influences the characteristics of cells. Still, the nucleus was not clearly implicated in inheritance—that is, in the contribution of parents to their offspring via gametes—until the cellular events involved in fertilization were described in the late 1870s.

Fertilization: Combining two sets of chromosomes

Even before fertilization was described, biologists had reason to suppose the nucleus was involved. Beginning in the middle of the eighteenth century, plant breeders had noted that in the inheritance of alternative traits, such as red or white flower colour, the sex of the parents did not usually influence the outcome of a cross. *Reciprocal crosses*—for example, a white male crossed with a red female and a white female crossed with a red male—produced identical results. Yet the egg in all species is much larger than the sperm; the volume of *cytoplasm*, or material outside the nucleus, is up to 1000 times greater in the egg. If the cytoplasm were responsible for inheritance, one would expect the genetic contribution of the female to be much greater than that of the male. All this remained rather speculative, however, as long as the mechanism of fertilization was unknown. Since many nineteenth-century researchers believed that more than one sperm is required to fertilize an egg, the argument about relative cytoplasmic volumes was not entirely convincing.

Working with the gametes, or sex cells, of plants and of sea urchins, researchers in the late 1870s observed that fertilization involves the union of a single egg with a single sperm and that the nuclei of the two gametes fuse shortly after the cells unite. They concluded correctly that fusion of the nuclei is the critical event in fertilization and that the fertilized egg's nucleus is composed of nuclear material from both parents. There was now impressive evidence that the chromatin is continuous not only from one cell generation to the next but also from one generation of multicellular organism to the next.

Such continuity made chromatin a promising candidate for the role of hereditary substance.

Chromosome pairs: A consequence of fertilization

A consequence of fertilization is that the zygote contains two sets of chromosomes, one from each parent. In animals, this means that every somatic cell also contains two sets of chromosomes, since every somatic cell is ultimately derived from the zygote by mitotic divisions, which reproduce the chromosomes exactly. (The situation in many plants and a few animals is more complicated.) We refer to a cell with two chromosome sets as a *diploid* cell and to its chromosome number as the diploid number, or 2n. Each chromosome in a diploid cell, with an exception we will discuss later, has a *homologous chromosome*, or *homolog*, very much like itself in appearance and genetic information, but derived from the other parent. However, this situation is not particularly evident in the mitotic nucleus, where the chromosomes are mixed together randomly and homologs may be far apart. Though cytologists realized after observing the fusion of sex cell nuclei that fertilized eggs contain chromosomes from both parents, they did not know that chromosomes come in pairs until they observed a special kind of cell division in which homologous chromosomes pair up with each other.

Meiosis: Halving the nuclear material

The fusion of gametes presents a logistical problem. If the nuclear material of two somatic cells is combined in the zygote, one would expect the quantity of nuclear material—the number of chromosomes—to double in every generation. It is clear that this does not happen, for every species has a characteristic chromosome number. Humans, for example, have a diploid number of 46 chromosomes, meaning that they have 46 chromosomes in most somatic cells. Cytologists therefore suspected that gametes must contain only half the amount of nuclear material found in somatic cells, and this idea was confirmed when *meiosis* (Greek *meion*, "smaller") was described in 1883.

Meiosis is the type of cell division that produces the gametes. (Again, we are speaking primarily of animals, for in plants meiosis may occur at a different stage in the life cycle.) In meiosis, a diploid cell undergoes two nuclear divisions after chromosome replication to produce four cells, each with half the usual number of chromosomes. Such cells are called *haploid*, and the haploid number of chromosomes is designated n, indicating a single set. Thus $n = 23$ in humans, and the diploid number, 46, $= 2n$.

In the first meiotic division, called meiosis I, the chromosomes are arranged in pairs, each alongside its homolog. Each chromosome is composed of two chromatids, having replicated prior to cell division. As the cell divides, one member of each chromosome pair goes to one of the newly forming nuclei and its homolog goes to the other, without any splitting of the centromeres Early cytologists suggested that all the chromosomes of maternal origin go to one nucleus and all those of paternal origin to the other, but, infact, it is purely a matter of chance which member of a given chromosome pair ends up in which nucleus: the chromosome pairs assort independently. At the end of meiosis I, each new nucleus contains half the original chromosomes—one of each pair— although the chromosomes still contain two identical chromatids. Thus, meiosis I is a *reduction division*.

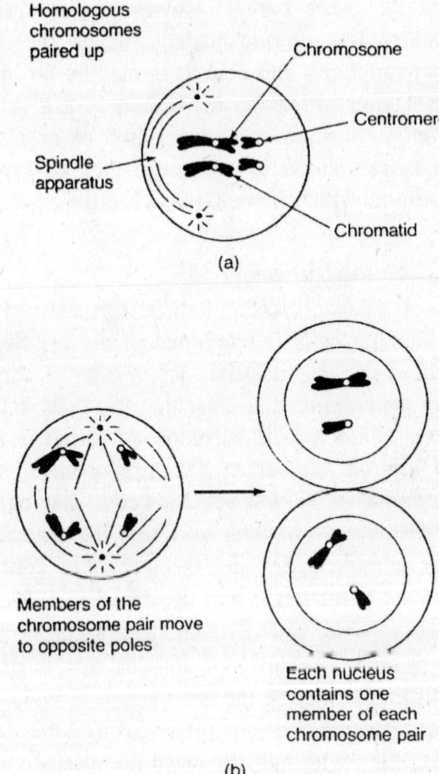

Fig. 11.4. A key feature of meiotic cell division is (a) the synthesis, or pairing, of homologous chromosomes and (b) the segregation of the pair members into different nuclei.

In meiosis II the nuclei divide again, but this time the centromere of each chromosome divides, and the two chromatids pull away from each other, each going to a new nucleus. Meiosis II is superficially like mitosis. At the end of meiosis, there are four haploid nuclei containing single-stranded chromosomes.

Whereas mitosis preserves the number of chromosomes in each cell, meiosis reduces the number to half the original. Fertilization then restores an organism's full chromosome complement the major differences between mitosis and meiosis, and illustrates the preservation of the chromosome number in a human life cycle. (Recognize that though the life cycle of plants is much more complex than that of animals, the critical event is the same: haploid gametes unite to form a diploid zygote. This means that the principles of chromosome transmission are the same in peas as in people.)

Details of Mitosis

The life history of somatic cells, called the *cell cycle*, consists of successive mitotic divisions (mitoses) separated by a period of non-division called the *interphase*, in which a cell replicates its genetic

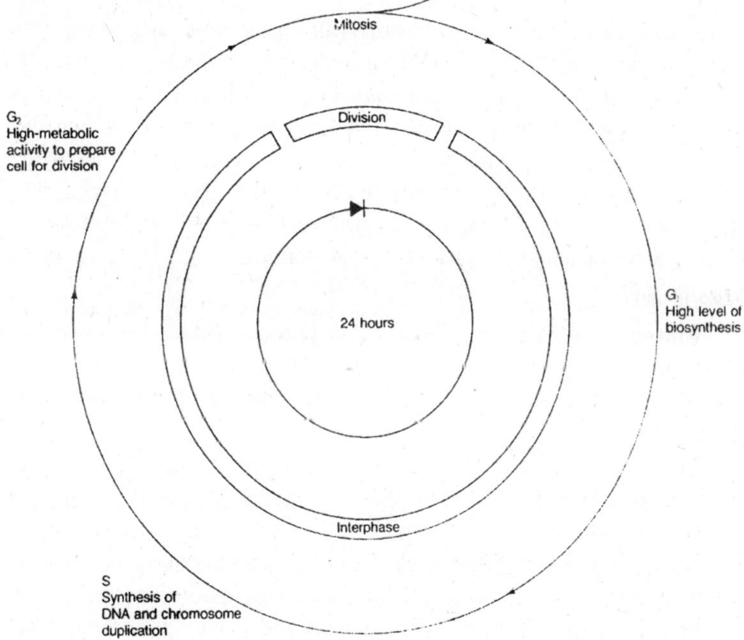

Fig. 11.5. The cell cycle.

material. Each interphase, then, is preceded and followed by a mitosis phase, sometimes called the *M phase*. The proportion of time the cell spends in each of these stages.

Interphase

The chromatin in the interphase nucleus is visible in high-magnification electron micrographs as long strands, jumbled together like strands of spaghetti in a bowl. Each chromatin strand is a highly uncoiled chromosome during mitosis it will coil tightly and condense into the chromosome form visible under the light microscope. The *nucleolus*, a nuclear entity whose constituents function in protein synthesis, is clearly visible through a light microscope during interphase.

Early cytologists sometimes called interphase the 'resting phase," but the term is quite inaccurate. The interphase cell is constantly engaged in its normal metabolic activities of molecular synthesis and energy production. During the interphase, the cell grows to approximately twice its original size. The chromatin, or rather, that part of the chromatin called DNA, directs the synthesis of proteins and is thus directly or indirectly responsible for all the cell's growth and activity.

Replication of the chromosomes also occurs in interphase. The period during which this takes place is designated the S (for synthesis) phase, and the periods preceding and following it are called G_1 and G_2 (for gap) phases. The length of G_1 is extremely variable. It may last anywhere from a few minutes to the lifetime of the organism, depending on how often the particular kind of cell divides. The G_2 is generally quite short. During the G_2 the chromosomes are still in the form of lengthy strands, but each strand is now double.

Mitotic division

Mitosis is divided into four main phases. These are somewhat artificial, since the process of nuclear division is actually continuous, but biologists have long divided the process into phases for convenience in description.

The first stage of division, called *prophase*, begins when the chromosomes start to coil and condense into structures visible through the light microscope. At the same time, a *spindle apparatus* begins to form in the cytoplasm. This structure consists of slender protein fibers called *spindle fibers* attached to two cytoplasmic bodies called *centrioles*; the spindle fibers will contract and move the chromosomes to opposite poles of the cell. By mid-prophase the chromosomes are

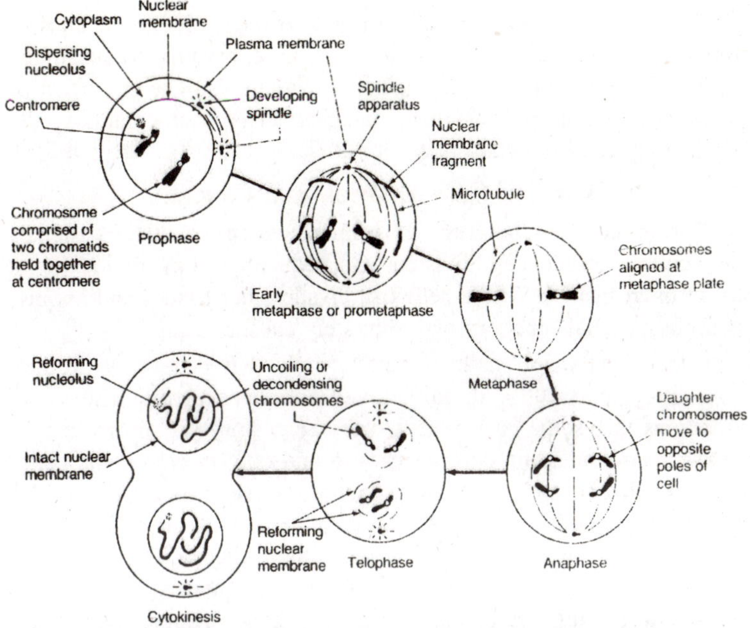

Fig. 11.6. The stages of mitotic cell division.

distinct and obviously double. The material of the nucleolus disperses throughout the nuclear region during prophase and is not visible through the light microscope for the rest of the cell division.

The second stages, *prometaphase* and *metaphase*, begin when the nuclear membrane breakdown and the nucleolus is completely dispersed. The chromosomes continue to coil and thicken, and become aligned randomly at the cell's equatorial region, or *metaphase plate*. Some of the spindle fibers are now seen to extend from the centrioles to the centromeres of the chromosomes, though others still extend from one centriole to the other.

The third stage, *anaphase*, begins when the centromeres separate as the *sister chromatids* of each chromosome move to opposite ends of the cell. After separation, sister chromatids are referred to as *daughter chromosomes*.

The final stage, *telophase*, occurs when the chromosomes have completed their movement toward the poles. It is marked by the formation of two daughter nuclei and two cells. Nuclear membranes form around the two groups of chromosomes; the chromosomes uncoil; the spindle apparatus disappears, and *cytokinesis*, or division of the cytoplasm, occurs. Narrowly, the term *mitosis* refers only to the division of the nuclear material, but it is often used broadly to mean the entire process of cell division, including cytokinesis. The end result of mitosis is two genetically identical daughter cells, each containing the diploid number of chromosomes.

Mitotic Cell Division and Cancer

Cancer cells do not respond properly to the regulatory signals that control mitotic cell division. Consequently, they divide in an uncontrolled fashion. These cells can invade other tissues and organs, disrupting normal functions and ultimately causing death.

Cancer cells are mature somatic cells that behave as if they were embryonic cells, dividing over and over again. There are indications that some cancer cells carry regulatory protein molecules in their membranes that are normally found only in embryonic cells. These regulatory molecules allow embryonic cells to divide rapidly for specific periods of time. In the absence of these regulators, cell division is controlled by a different regulatory scheme. The regulatory proteins that trigger the rapid cell division are normal structurally, but they are being synthesized in the cell at an inappropriate time. Cancer cells are normal in most respects, except that some of their genes are acting in the wrong place at the wrong time to alter the regulatory scheme of mitosis.

What causes a normal cell, undergoing a normal mitotic cycle, to become cancerous? We shall explore this question in more detail later, but one cause of cancer may lie with certain types of viruses. These viruses may be able to change the pattern of gene activity in a mature cell so that a gene normally active only in the embryo becomes activated. Transforming the cell into a cancer cell.

Details of Meiosis

Meiosis is a lengthy process in which the chromosomes duplicate once and the nucleus divides twice, thus reducing the chromosome number by half. Like the term *mitosis*, the term *meiosis* may refer to the nuclear division alone, or to nuclear division an the accompanying cytokinesis. Meiosis result in the formation of four haploid cells from one diploid cell.

Each of the two nuclear divisions of meiosis has phases with the same names as those of mitosis. However, the phases of meiosis I and the corresponding phases of mitosis bear only a superficial resemblance to each other.

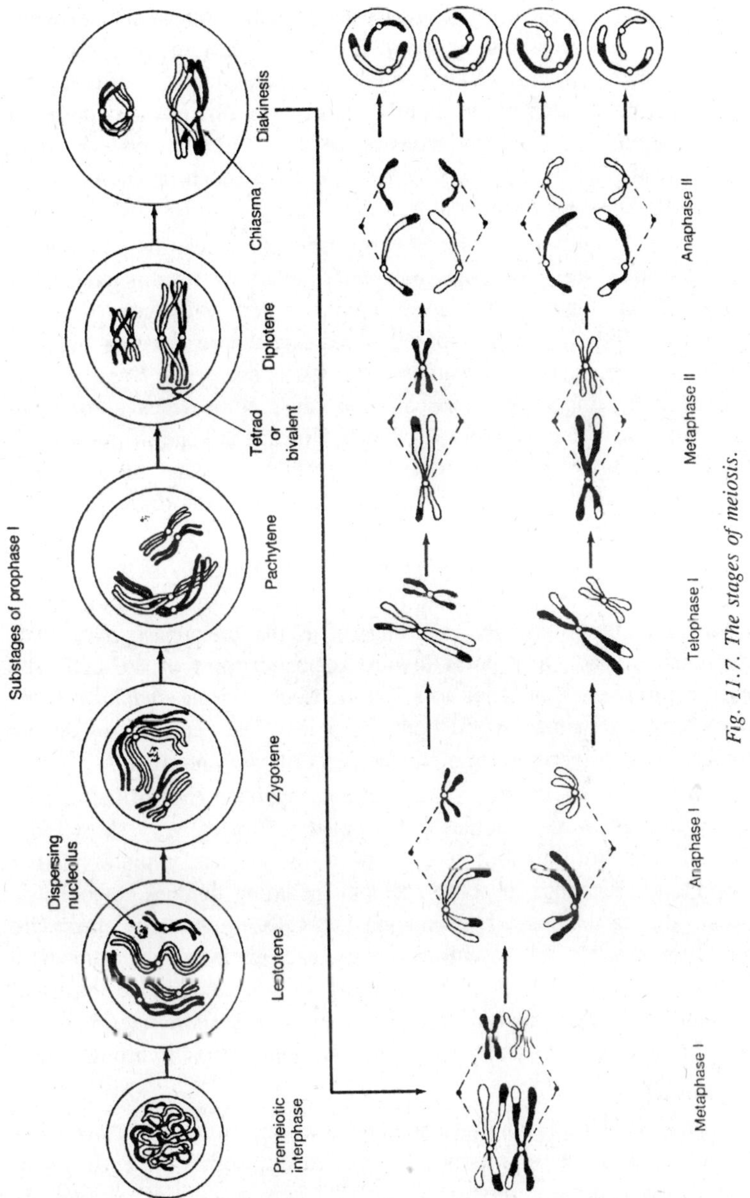

Fig. 11.7. The stages of meiosis.

Meiosis I

The chromosomes replicate during the interphase preceding meiosis I. In *prophase I*, the first stage of meiosis I, each double-stranded chromosome comes to lie alongside its homolog and then intertwines with it in a process called *synapsis*. At the start of synapsis, each chromosome pair looks like a single structure, but as the homologs pull away from each other slightly, it becomes obvious that there are two of them and that each consists of two sister chromatids. Each intertwined pair is called a *bivalent* (for the two chromosomes) or a *tetrad* (for the four chromatids).

As the lengthy and complex prophase I continues, the homologous chromosomes seem to repel each other, but they remain attached at some points, forming X-shaped regions called *chiasmata* (singular: *chiasma*). The chiasmata apparently indicate where *crossing over*, or exchange of homologous chromosome parts, is occurring. Crossing over is extremely important genetically be cause it provides a basis for generating new combinations of alleles. Toward the end of prophase I, the bivalents start moving toward the metaphase plate, the nucleolus begins to disperse, and a spindle apparatus starts to form in the cytoplasm.

Meiotic metaphase I begins with the disassembly of the nuclear membrane. During metaphase I, the bivalents, their members still attached at the chiasmata, are aligned at the metaphase plate. The members of each pair point toward opposite poles of the cell, and each centromere is attached to a spindle fiber. During *anaphase I*, the homologs are pulled toward opposite poles. The centromeres do not divide, so each chromosome still consists of two chromatids.

When the chromosomes have reached opposite ends of the spindle apparatus, *telophase I* begins. This phase is usually very short, with the chromosomes uncoiling only partially and incomplete nuclear membranes forming. In some species, including humans, cytokinesis occurs during telophase I, producing two cells. In any case, there are now two separate sets of chromosomes, each with half the number in the parental cell, but each chromosome still contains twice the usual amount of genetic material. There is only a very short interphase, or none at all, after meiosis I, and there is no duplication of chromosomes.

Meiosis II

Meiosis II superficially resembles a normal mitosis. In *prophase II*, the first phase of meiosis II, the chromosomes in each group recondense and start moving toward the new equatorial regions. At *metaphase II*, the chromosomes are aligned at the metaphase plates, with spindle fibers attached to their centromeres. During *anaphase II*, sister chromatids are pulled apart (the centromeres split), and the daughter chromosomes move to opposite poles. *Telophase II* begins when the chromosomes reach the poles and cease movement. Nuclear membranes now form around each haploid complement of chromosomes, and cytokinesis occurs. The second division completed, we have four haploid cells with a single set of chromosomes.

Autosomes and Sex Chromosomes

We mentioned earlier that there is an exception to the rule that every chromosome in a human somatic cell has a similar homolog. This exception is found in the cells of males. In 22 of a male's chromosome pairs the homologs are indeed very like each other, but the twenty-third pair consists of one chromosome called X and one very small one called Y. The X and Y chromosomes are not homologous and are not paired along their lengths in meiosis, but have an end-to-end association. The corresponding pair of chromosomes in a female consists of two homologous X chromosomes. The X and Y chromosomes are called *sex chromosomes*, and the other 44 chromosomes, which are the same in both sexes, are called *autosomes*.

It has been known for a long time that the X and Y chromosomes determine a person's sex. When gametes are formed by meiosis, the two sex chromosomes go to different cells, just like the members of every other chromosome pair. This means that each of a female's eggs will have an X chromosome, but half of a male's sperm cells

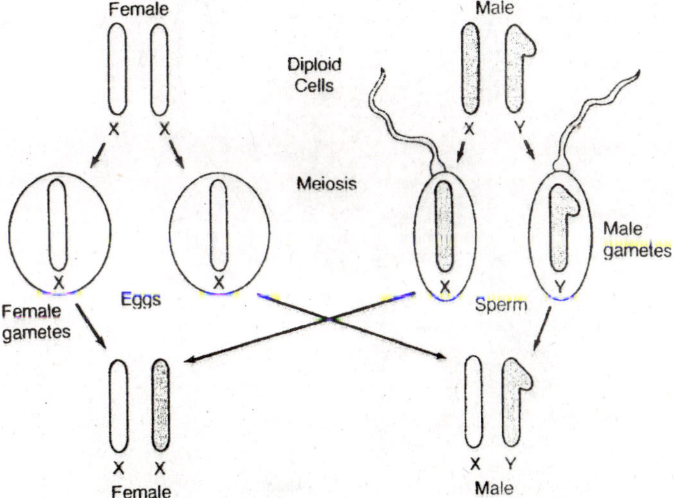

Fig. 11.8. Segregation of the sex chromosomes at meiosis results in a single kind of egg but two classes of sperm.

will contain an X and the other half a Y. At fertilization, a zygote normally receives an X chromosome from the mother and either an X or a Y from the father and will develop accordingly as a female or a male. The formation of equal numbers of X- and Y-bearing sperm is the reason for the approximately equal number of female and male births.

Gamete Formation arid Fertilization in Humans

In humans and other mammals, meiosis occurs in the gonads—the male testes and the female ovaries—and produces sperm and ova, or eggs. The process of gamete formation is called *spermatogenesis* in the male and *oogenesis* (oh-oh-genesis) in the female. We will consider spermatogenesis first, since it is the simpler of the two processes.

Spermatogenesis

In sexually mature human males, sperm cells from continuously in the seminiferous tubules of the testes. The cells that line these tubules and ultimately give rise to the sperm are called *spermatogonia*. Spermatogonia go through several mitotic divisions over a period of three to four weeks, the final division producing cells called *primary*

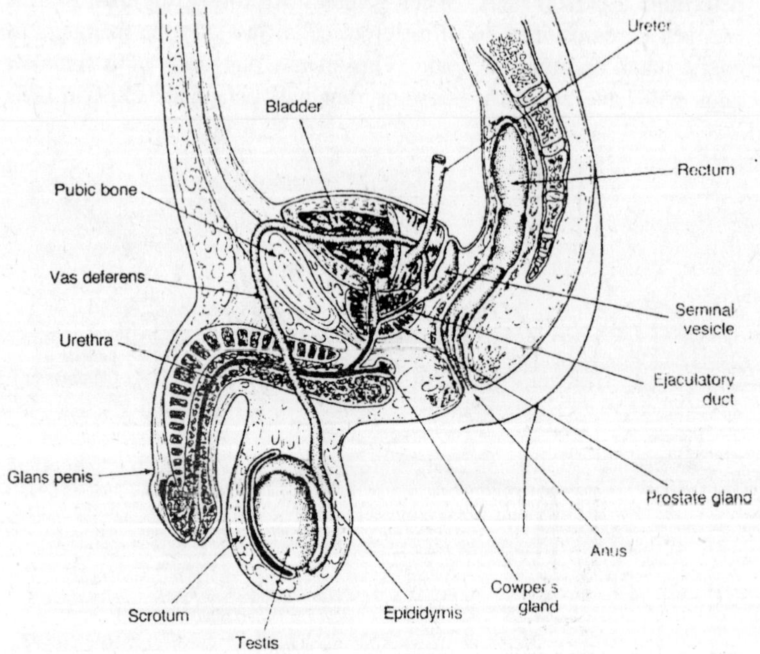

Fig. 11.9. Male reproductive system.

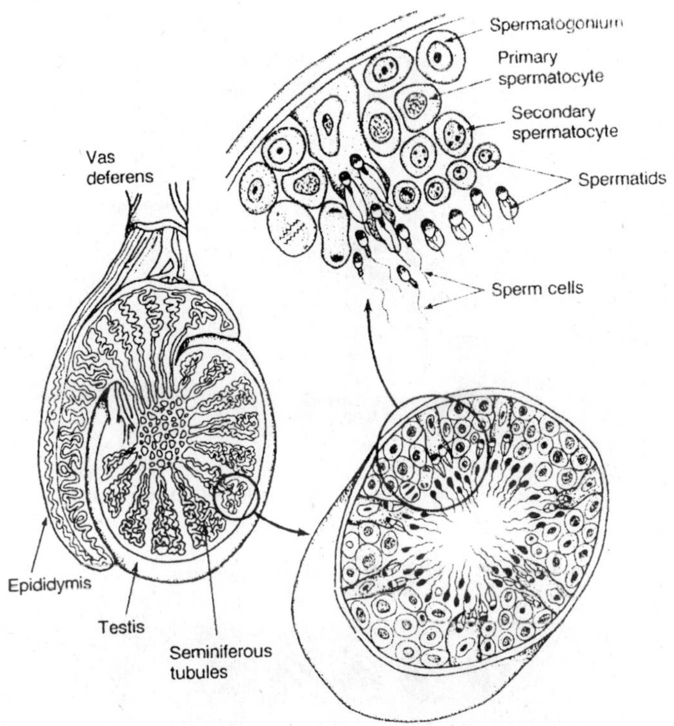

Fig. 11.10. Arrangement of seminiferous tubules in the testes, with a cross section of a seminiferous tubule and a closer view of the cell undergoing spermatogenesis.

spermatocytes. These diploid cells undergo meiosis I, producing *secondary spermatocytes*, each containing 23 double-stranded chromosomes. The secondary spermatocytes undergo meiosis II to form haploid *spermatids*, each with 23 single-stranded chromosomes.

The newly formed spermatids have an almost normal amount of cytoplasm and remain connected to each other by a cytoplasmic bridge. This allows all the sperm, both the X and Y forms, to share a common cytoplasm. One major function of this bridge is to ensure that sperm with a Y chromosome have access to gene products coded for by genes on the X. The X chromosome carries many essential genes that are not found on the Y. Without the cytoplasmic bridge, the Y-bearing sperm would not survive. As the spermatids mature into spermatozoa, a process that takes from 10 hours to 16 days, they lose nearly all their cytoplasm and develop a tail for swimming. The *mature sperm* is essentially a nucleus with a tail—a packet of genetic material capable of moving about.

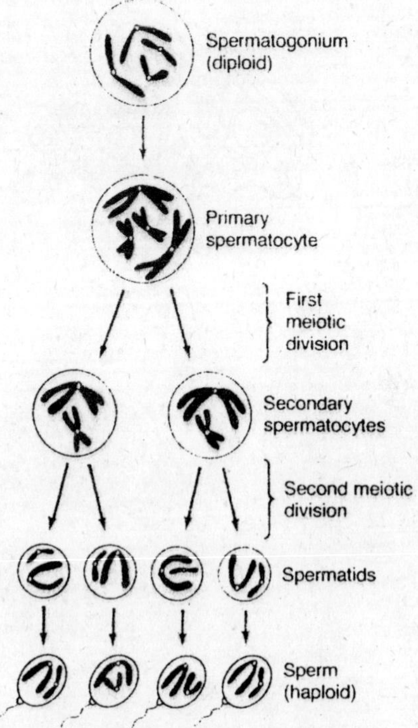

Fig. 11.11. The meiosis of spermatogenesis.

Mature sperm are stored in the epididymis (a cordlike structure at the back of the testis) until ejaculation, when they mix with secretions of the seminal vesicles and prostate gland to form *semen*. This mixing takes place in the latter portions of the vas deferens, or sperm duct, and the semen is ejected via the urethra. Three hundred million sperm cells may be present in a single ejaculate, about 100 million per cubic centimeter of fluid. When these sperm are deposited in the vagina, they move quickly into the cervical canal, but only a few dozen of them manage to make the long trip to the upper part of the fallopian tube, or oviduct, where an egg may be waiting. Sperm are viable for 48 hours or so after ejaculation, so if an egg is released into the fallopian tube within roughly two days after intercourse, fertilization may still occur. Once a sperm and egg unite, all other sperm are excluded.

Even though only one sperm is needed to fertilize an ovum, large numbers of active sperm are required to produce the enzymes that allow a sperm cell to fertilize the ovum. If a man's sperm count falls

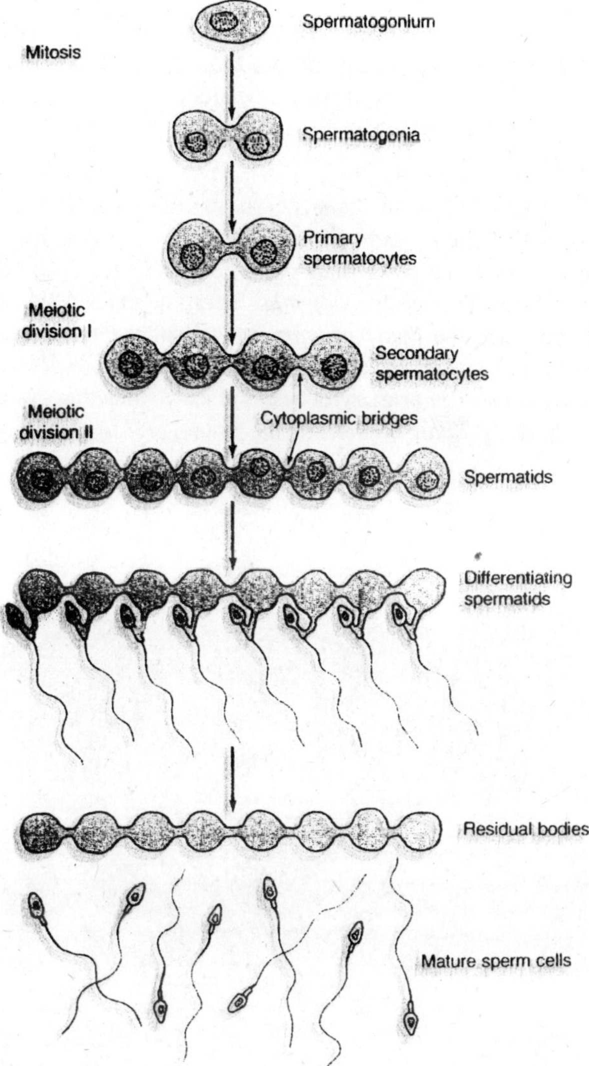

Fig. 11.12. Diagram showing how the progeny of a single spermatogonium remain connected to each other by cytoplasmic bridges throughout their differentiation into mature sperm.

much below 20 million per ejaculation, he will probably be sterile. A number of factors can interfere with sperm development, resulting in infertility or even birth defects. Cell metabolism in the male reproductive tract is very high, and this causes chemicals to be concentrated in the semen. The seminal fluid is a very sensitive indicator of toxins. Lead, tobacco products, THC from marijuana, large amounts of alcohol, and many pesticides may be concentrated in semifinal fluid and result in lower sperm counts.

Oogenesis and fertilization

Oogenesis in humans begins early in embryonic life, between the eighth and twentieth weeks of development. Cells destined to become female gametes, the *primordial germ cells*, migrate to the developing ovary, where they become *oogonia*. These oogonia divide rapidly by mitosis to form the many *primary oocytes* of the female fetus. Each oocyte is surrounded by a layer of specialized secretory cells called *granulosa cells*. The primary oocyte and its surrounding granulosa cells constitute the *primary follicle* of the fetus.

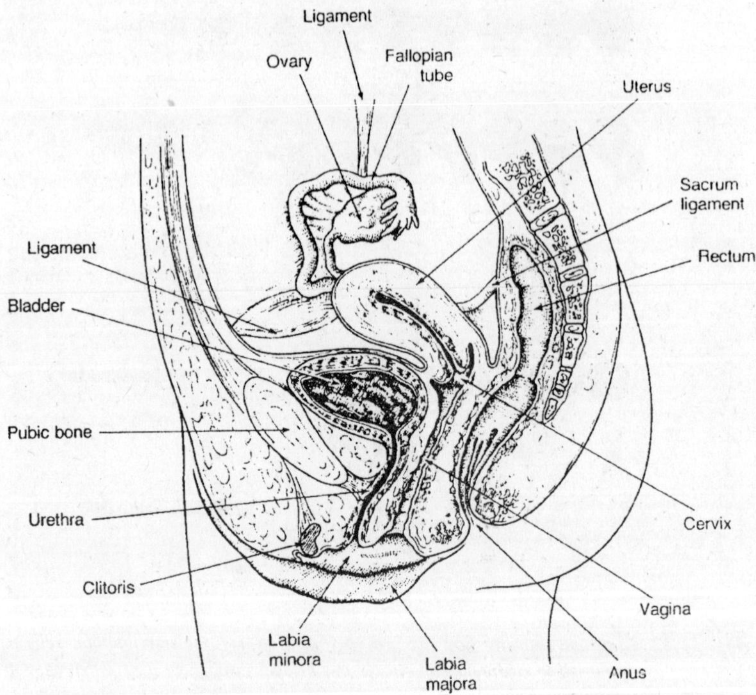

Fig. 11.13. Female reproductive system.

Unlike meiosis in males, meiosis in females begins in the third month of intrauterine life. The primary oocytes enter the long prophase of the first meiotic division. The DNA has replicated so that each chromosome consist of two chormatids; the homologous chromosome

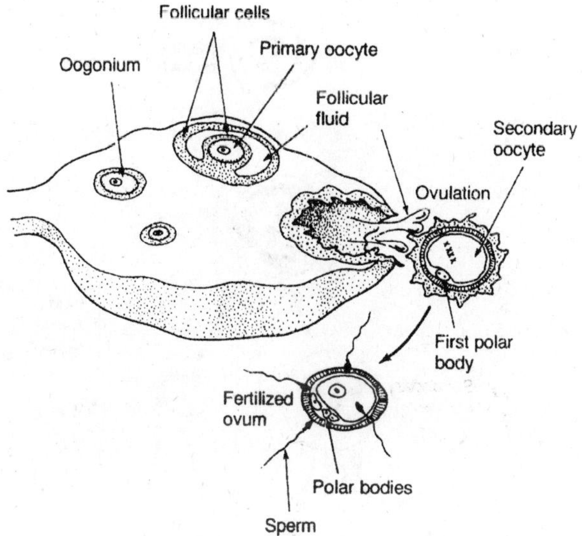

Fig. 11.14. Cross section of an ovary, showing the development of an ovum (egg cell) from a primary oocyte.

are paired and crossing over occurs between the chromatids or homologous pairs. At this point everything stops. The primary oocytes remain in a state of suspended prophase I until the female reaches puberty. Thus, unlike the male, the female is born with all her potential gametes already in a state of meiosis.

Beginning at puberty and continuing for the reproductive period of a woman's life, about 40 years, one egg a month re-enters the meiotic process. Under the direction of specific hormones, the replicated chromosomes segregate into two daughter nuclei; each containing half the original number of chromosomes however the cytoplasm does not divide equally one daughter cell the *secondary oocyte*, get essentially all of the cytoplasm while the other, called a *polar body*, gets very little. At ovulation, the secondary oocyte breaks out of the follicle and moves into the fallopian tube, with the polar body attached to it. Fertilization normally takes place in the fallopian tube within a day or so of ovulation.

The second phase of meiosis is completed in the oviduct only if the ovum (secondary oocyte) has been fertilized. Meiosis II is very rapid. The centromeres separate and daughter cells appear, but again there is an Unequal distribution of cytoplasm. The secondary oocyte divides into a *mature ovum* and a second polar body. The first polar body also completes meiosis II by producing two very small polar

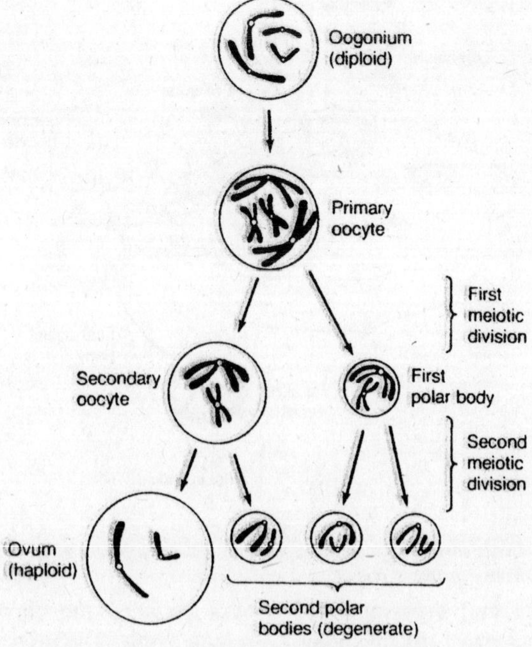

Fig. 11.15. The meiosis of oogenesis.

bodies. The mature ovum is about 1000 times the volume of a polar body (or a sperm, for that matter).

In contrast to meiosis in the male, which results in four functional sperm, the result of meiosis in the female is four haploid cells, but only one of them is a functional gamete. The three polar bodies are never fertilized, as far as we know, and simply disintegrate within the fallopian tube.

The nucleus of the mature ovum fuses with the sperm nucleus about one hour after fertilization to form the diploid *zygote*. As the zygote travels through the fallopian tube, it undergoes repeated mitotic divisions to form an embryo. Arriving in the uterus about five days after ovulation, the embryo becomes implanted in the thickened uterine lining, which supplies it with nourishment.

If an egg is not fertilized within about 24 hours of ovulation, meiosis does not go to completion, and the egg degenerates. After the egg has been released, if it has not been fertilized, the menstrual period occurs. This bleeding is caused by the sloughing off of the thickened uterine lining. A new lining starts to form as another primary follicle begins to mature. The *menstrual cycle*—the reproductive events from one menstrual period to the next is under hormonal control and takes an average of 28 days, its exact length varying from one woman to another.

Oogenesis appears to be a very wasteful process. Perhaps 1500 primordial germ cells invade the developing ovary and then undergo mitotic division to form about 7 million oogonia. Most of these oogonia degenerate and so never become primary oocytes. At birth there may be as many as 2 million primary oocytes left in the two ovaries. This degeneration process continues, and by puberty only about 250,000 primary oocytes remain. By menopause, only a few are left. During a woman's reproductive life, about 500 eggs are released. All the rest degenerate. We do not know why so many eggs are formed only to degenerate.

The Down syndrome case discussed at the beginning of this chapter is probably tied to the process of oogenesis. As a woman ages, say into her thirties and forties, the eggs she releases in her menstrual cycles have been in a state of suspended prophase for 30 or 40 years. The eggs can be damaged during this long wait, and if they are, the distribution of chromosomes to daughter cells during meiosis can be disrupted. This may explain why a 40-year-old woman has a 1% chance of having a child with Down syndrome, whereas a 20-year-old has only a 0.04% chance, a 25-fold difference.

Mating game: Sperm meets egg

The primary function of the mating game is to promote the union of gametes, which creates a new individual with a unique combination of genes. Many unicellular organisms reproduce by an asexual act of simple fission or mitosis, but in asexual reproduction the offspring are identical to the parents except when a *mutation*, or change in the structure of a gene, arises. It is the segregation and independent assortment of chromosomes during meiosis, followed by the random combination of sperm and egg that gives us our biologic variability. This scrambling of genetic material in every generation is called *genetic recombination*. The endless variation generated by recombination has provided an enormous evolutionary advantage to sexually reproducing species, since it allows for faster adaptation to a variety of conditions.

Human Chromosome

Viewing Human Chromosomes

Most early knowledge about chromosomes was derived from studies of nonhuman cells. Geneticists struggled for decades to develop techniques that would give them a reliable view of human chromosomes, but they encountered enormous difficulties. It was not easy to obtain human tissue to study; fixing and staining procedures produced erratic results; and it was not at all unusual to get contradictory data from seemingly identical experiments. You can appreciate the magnitude of the problems involved if you realize that until the mid-1950s geneticists did not even know the exact number of human chromosomes. Early investigators reported human chromosome complements ranging from 8 to 73, but in 1923 an authoritative study seemed to show that the normal diploid number is 48. This became the generally accepted number until 1956, when Joe-Hin Tjio and Albert Levan announced their use of new techniques to show that the correct number is 46. In retrospect, it seems amazing that the answer to so simple a question as How many chromosomes does a person have?" was not known until three years, after the structure of DNA was announced.

Tjio and Levan are to human cytogenetics what Mendel was to genetics. Their brilliant and painstaking improvements of existing methods for studying cells include two techniques that were particularly important to the success of their chromosome-counting work and that have since had widespread application in the study of human chromosomes. The first of these is the technique of placing dividing cells in hypotonic solutions (a technique that Levan learned from T. C. Hsu)—salt solutions of lower concentration than that of the cell contents. One reason for the earlier discrepancies in chromosome counts was the tendency of the chromosomes to overlap and stick together. Placing a cell in hypotonic solution causes water to flow inward, so that the cell swells up and the chromosomes separate from each other, becoming more easily distinguishable. The second technique that Tijo and Levan perfected was that of culturing mammalian cells. They grew lung fibroblast cells (immature connective-tissue cells) in a culture medium to obtain their material for chromosome study.

For their study of human cells. Tjio and Levan also adapted important techniques used in cytologic studies of other organisms. One such technique is the use of a stain called *orcein*, which stains chromosomes in a very distinct fashion, so that they are visible in more detail than with older staining procedures. An other was the arresting of dividing cells at metaphase by means of a drug called colchicine, derived from the autumn crocus. Colchicine interferes with the attachment of spindle fibers to the centromere, so that when chromosomes are arranged at the metaphase plate they cannot migrate

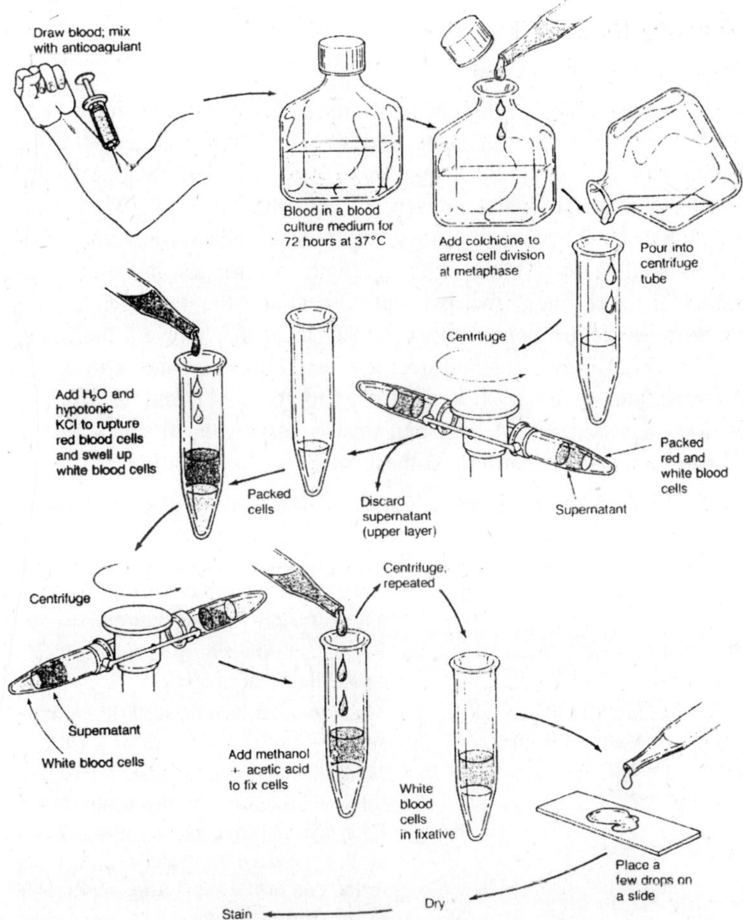

Fig. 11.16. The most common method of harvesting human chromosomes from lymphocytes.

to opposite poles of the cell. The use of colchicine stops mitosis and allows a large number of dividing cells to accumulate for observation.

Another great advance in human cytogenetics occurred when Moorehead developed a method of culturing lymphocytes (a class of white blood cells) taken from the circulating peripheral blood (i.e., blood near the surface of the body). Obtaining blood samples is vastly easier than obtaining samples of other human tissues, so the problem of material for study was at least partially solved by this technique, which is now standard laboratory procedure. The most common method of harvesting human chromosomes from lymphocytes is outlined.

Identifying Human Chromosomes

Chromosome classification

In recent years, cytogeneticists have devoted major efforts to establishing uniform standards and nomenclature (names) for the classification of human chromosomes. Toward this end they have held a series of international conferences beginning in 1960. They have agreed that the 22 pairs of autosomes (non sex chromosomes) should be numbered in descending order of length—the longest has the lowest number, 1—and that if two pairs are essentially the same length, the pair with the centromeres most centrally located would get the lower number. They have further decided that chromosomes should be organized into seven groups according to their size and centromere position. A schematic of a human male's *karyotype*, or chromosome makeup, arranged according to these rules of classification.

Table 11.1. A general classification of the human chromosomes

Group	*Chromosomes*	*Characteristic*
A	1, 2, 3	Large, metacentric chromosomes'; approximately median centromere.
B	4, 5	Large, submetacentric chromosomes"; submedian centromeres.
C	6 through 12 and X (sex chromosome)	Medium-sized, submetacentric chromosomes.
D	13, 14, 15	Medium-sized, acrocentric" chromosomes with nearly terminal centromeres. Each chromosome in this group has a small appendage, or "satellite," attached to the end of the short arm, observable in good preparations
E	16, 17, 18	Shorter than group D; metacentric or submetacensric
F	19, 20	Short; metacentric
G	21, 22, Y (sex chromosome)	Very short, acrocentric chromosomes; 21 and 22 have small satellites. Again, these satellites are observed only in better preparations

In 1960, when these classifications were being worked out, only a few chromosome pairs (1, 3, and 16) could be identified unambiguously on the basis of length and centromere position. The discovery of additional physical landmarks, such as secondary constrictions (the primary constriction is the one at the centromere), enabled

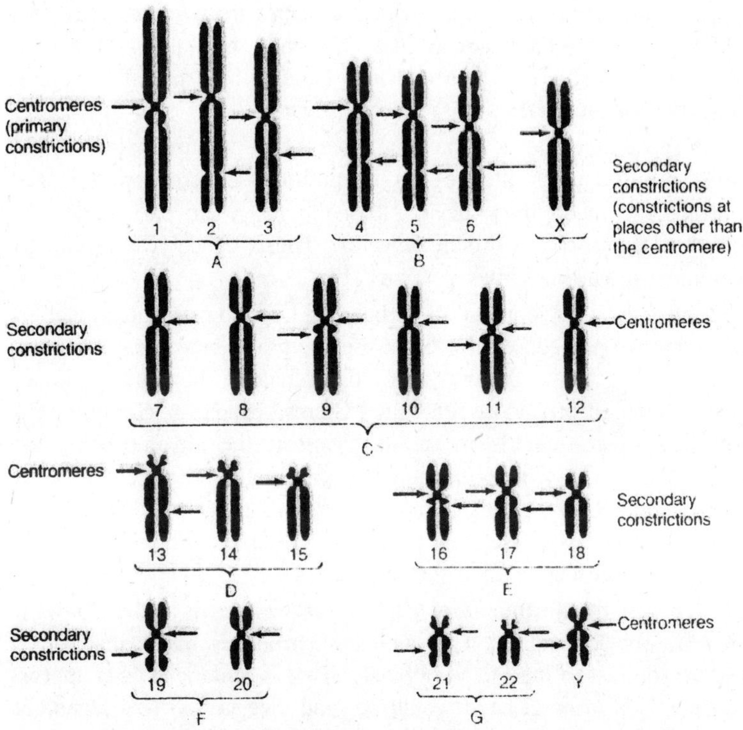

Fig. 11.17. Schematic of the normal male karyotype (chromosome makeup).

cytogeneticists to identify other chromosomes. However, a true revolution in human cytogenetics occurred when banding techniques were developed.

Chromosome banding

During the late 1960s and early 1970s, several new staining procedures were discovered that gave the chromosomes truly distinctive appearances. These procedures caused various regions of the chromosomes to stain differently, prompting the appearance of light and dark *bands*. Banding patterns are the most powerful tool now available for chromosome identification, because the pattern for every chromosome pair is unique. Furthermore, different regions of the same chromosome can be consistently identified by their banding patterns.

The first of the major banding techniques to be discovered uses quinacrine mustard, a fluorescing dye that preferentially binds to specific regions of chromosomes. When chromosomes treated with this dye are viewed under ultraviolet light, fluorescent bands of varying intensity appear. The pattern is characteristic for each chromosome. These are called *Q bands* (Q for quinacrine). The most brilliantly fluorescing chromosome is the Y, which makes Q-banding an especially valuable technique for identifying the Y chromosome.

A second technique, C-banding, uses a dye mixture called Giemsa stain, which stains the region around the centromere of each chromosome much more deeply than the rest. The dark bands are called *C bands* (C for centromeric). These bands are especially prominent in chromosomes 1, 9, and 16.

After the development of C-banding, cytologists found that by modifying the procedure, they could use Giemsa stain to show a banding pattern similar to the Q bands, but more detailed. The bands produced by the modified procedure are called Giemsa bands, or *G bands*. The G-banding technique is most successful if the chromosomes are pretreated with a protein-digesting enzyme called trypsin before the stain is applied. G banding is more common than Q-banding be cause it is less expensive, yields a permanent slide record, and does not require a fluorescence microscope.

The last of the important banding procedures is called "*reverse Giemsa*," or R-banding. This technique produces bands that are in reverse contrast to the Q and G bands. That is, a dark *R band* appears where a light Q or G band would be, and vice versa. This technique is especially valuable for observing the ends of chromosomes, which come out very dark.

Banding revealed such intricate details of each chromosome that new standards had to be set. So at the Paris conference of 1971, cytogeneticists established a system of classification for the chromosome bands and regions. This logical system uses numbers and letters to enable us to pinpoint any band in which we might be interested. The short arm of each chromosome is called the *p arm* ("p" for French *petite*, "small"), and the long arm is the *q arm*. Regions and bands are numbered from the centromere out. To identify a band, a sequence of four items is used: chromosome number, arm, region, and band number. For example, 9q34 refers to chromosome 9 the long arm, region 3, band 4. This band is indicated with an arrow in. For the sex chromosomes, an X or a Y is used instead of the first number.

Chromosome Structure

As soon as the chromosome was identified as the carrier of genetic information, scientists began to seek in its chemistry and structure an answer to the question "What is a gene?" Interestingly enough, they had a satisfactory answer to the question before they had a completely satisfactory picture of a chromosome for the answer lay in the molecular structure of one constituent of chromosomes—the *DNA*.

In the early years of this century, chemical analysis showed that chromosomes contain DNA (deoxyribonucleic acid) and protein. In the 1940s, Avery, MacLeod and McCarty performed experiments in which bacteria were genetically transformed by assimilation of DNA isolated from different strains. This demonstrated that the DNA, not the protein, was the genetically significant part of the chromatin strand. This was confirmed in 1952 when Hershey and Chase showed that in T4 bacterial viruses, DNA, not protein, was the genetic material. With the exception of certain viruses, DNA is confirmed as the genetic material in all living things. In 1953, Watson and Crick published their famous paper on the molecular structure of DNA, and their model, which has remained virtually unchanged for 30 years, quickly proved to hold the key to the genetic code. Only in the 1970s, however, did a clear picture begin to emerge of how DNA and protein are organized in the chromosome.

Genes are DNA

Shortly after the turn of the century, some biologists proposed that the chromosome is an inert "home" structure for genes, which journey away and perform their functions in other parts of the cell. But we now know that genes are segments of chromosomal DNA, and that they direct the synthesis of proteins by means of intermediary molecules (called *ribonucleic acid*, or RNA) that are copied from DNA templates and then move into the cytoplasm through the pores of the nuclear membrane. A later view portrayed chromosomes as consisting of genes strung together like beads on a string. In some ways, the beads-on-a-string model is not entirely in accurate. We find that chromosomes consist of DNA and protein elaborately wound together in a complex three dimensional structure with genes arranged linearly. This linear view of gene organization was useful, for it allowed geneticists to work out the sequence of specific genes on chromosomes long before they understood the structure or functioning of genes.

Nucleosome model

the current view of mammalian chromosome structure. The basic molecular components of the chromatin strand are DNA and several varieties of a class of protein called *histories*. The DNA molecule consists of two spiral strands, linked together at intervals by hydrogen bonds to form a double helix. The chemical structure of the DNA double helix is the basis of the genetic code. A chromosome contains a continuous length of DNA, wound around individual structures made up of histones, rather like a continuous thread winding around a series of spools. Each DNA-histone unit is called a *nucleosome*. The nucleosomes coil and condense into units that appear as bands or chromomeres when they are appropriately stained. Further coiling and condensation produce the chromosome we are able to view with the electron microscope.

12

MITOSIS

There are two types of organisms-acellular and multicellular. The growth and development of an individual depends exclusively on the growth and multiplication of the cells. It was *Virchow* who first of all adequately stated the cell division. In animal cell the cell division was studied in the form of segmentation division or cleavage by *Prevost* and *Dumas* in 1824. The mechanism of cell division was not precisely investigated until long afterward but *Remak* and *Kolliker* showed that the process involves a division of both the nucleus and the cytoplasm. The term *karyokinesis* was introduced by *Schleicher* (1878) to designate the changes of nucleus during division, and the term *cytokinesis* was introduced by *Whiterman* (1887) to designate the associated changes taking place in the cytoplasm.

Cell division is necessarily the avoidance of aging, and secondly for the segregation of an individual into semi-independent units which leads to efficiency. Thus, we see that cell division is a widespread phenomenon that is essential not only for the maintenance of life but also for the development of the organism itself.

Cell division can be conveniently described as:

(i) *Direct division.* Where the nucleus and cell body undergo a simple mass division into two parts. It is also called *amitosis*.

(ii) *Indirect division.* Here the nucleus undergoes complicated changes before it is divided into two daughter nuclei.

AMITOSIS

The amitosis or direct cell division is the means of asexual reproduction in acellular organisms like bacteria and protozoans and

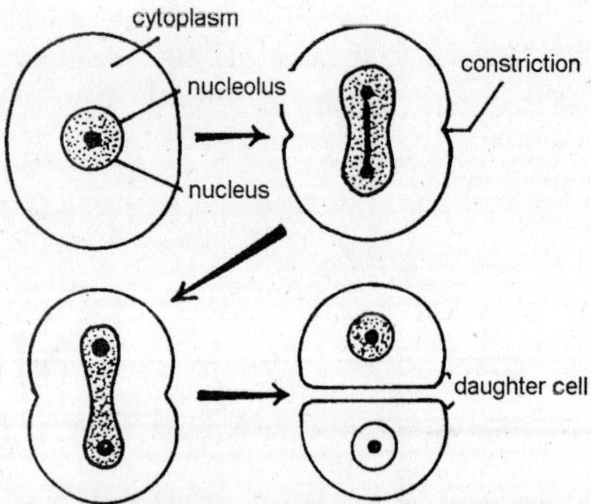

Fig. 12.1. Diagram to illustrate process of amitosis.

also a method of multiplication or growth in foetal membranes of some vertebrates. In amitosis type of cell division the splitting of nucleus is followed by cytoplasmic constriction.

During amitosis the nucleus elongates first and then assumes dumbell-shaped appearance. The depression or construction increases in size and ultimately divides the nucleus into two nuclei. The division of nucleus is followed by the constriction of cytoplasm which divides the cell into two equal or approximately similar halves. Therefore, without the occurrence of any nuclear event two daughter cells are formed.

Mitosis

In mitosis, one cell divides into two, both of which are genetically identical to each other and to the parental cell. In other words, both, the chromosomes and the genes, are the same in all of the cells. This type of cell division is necessary if the organism and/or cell is to persist and survive. There are many facts to the need for cell division, and they vary depending on the particular biological function. For example, in liver tissue when some cells die or are damaged, others divide and provide new cell to replenish those that are lost. Other cells in the organism actually grow (increase in size), and it is possible that when they reach a point where there is too much cytoplasm far a given amount of nuclear material, they divide, and the whole process begins again. The phenomena of growth also involves an increase in numbers of cells. The increase in size of a tissue or organ is often due to numerical increase in cells rather than an increase in cell size. When subjected to the appropriate environmental and biochemical signals, these cells may be stimulated to differentiate upto a specific cell type. The sum total is that, as a result of all division, a certain degree of plasticity and immortality is provided to the organism. When the plasticity is lost, the organism undergoes the process of aging yet when the process of division is out of control, the organism literally "grows" to death!

The process of mitosis is a continuous one but on the basis of certain changes the same can be divided into several phases or stages.

Characteristic Features of the Mitotic Phases or Stages

In cell division, the first visible changes occur inthe nucleus, when the chromonemata condense into *chromosomes* (Gr. coloured bodies). This stage is called the *Prophase* (Gr. early figure).

Then the nuclear boundary disappears and the chromosomes line up in or near one plane in the cell. This is *Metaphase* (Gr. middle figure).

Following this each chromosome separates into two parts, and these two parts migrate away from each other to the ends of the cells. This is *Anaphase* (Gr. up and down figure.).

Next the chromosomes at each end of the cell reconstitute a nucleus. This is *Telophase* (Gr. final figure).

Then the cytoplasm divides into two daughter cells, bringing about the *Interphase* again.

Interphase

The period of metabolic activity during which cell division is not in a process, has been called the 'interphase.' This is frequently referred to as the 'resting phase' but this term is not appropriate because the cell is metabolically most active at this stage that is why *Berril* and *Huskins* (1936) referred it as the *energy phases*. The interphase is the period between the telophase of one division and the prophase of the new cell division. During this phase the cell does everything except division. It is during interphase that the genes self-duplicate and carry on their function of supervising synthesis.

The chromatin granules in the nucleus are not readily distinguishable in the living cell, but may be brought out by treatment with chemicals which kill, fix and stain them. They appear at first glance to be scattered throughout the nucleus, but careful study has produced evidence that they are arranged in a definite pattern as long coiled strands and appear as a thread like net work in ordinary stained preparation.

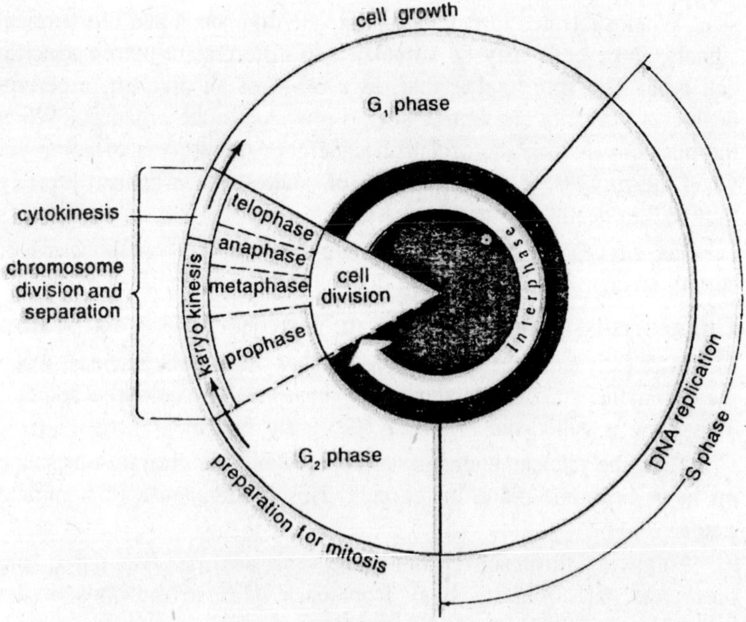

Fig. 12.2. Stage of the cell cycle.

"Nuclear sap" or, "Karyoplasm" fills the interstics between the chromosomes. One of more rounded bodies, the *nucleoli* are usually present. Between the nucleus and the surrounding cytoplasm, is the *nuclear membrane*. In the cytoplasm adjacent to the nucleus, there is a body, the *central body*, which consists of two granules or, after each granule has replicated, of two pairs of granules, the *centrioles*.

A typical cell cycle, including interphase lasts from 20-24 hrs. Interphase in the longest period in the cell cycle, and may last for several days in cells.

Interphase can be further divided into form sub-phases:

1. G_1-phase
2. S-phase
3. G_2-phase
4. M-phase.

G_1-phase includes the synthesis and organization of the substrate and enzyme necessary for DNA synthesis. Therefore, G_1, is marked by the synthesis of RNA and protein.

G_1-phase is followed by the S-phase where the synthesis of DNA occurs.

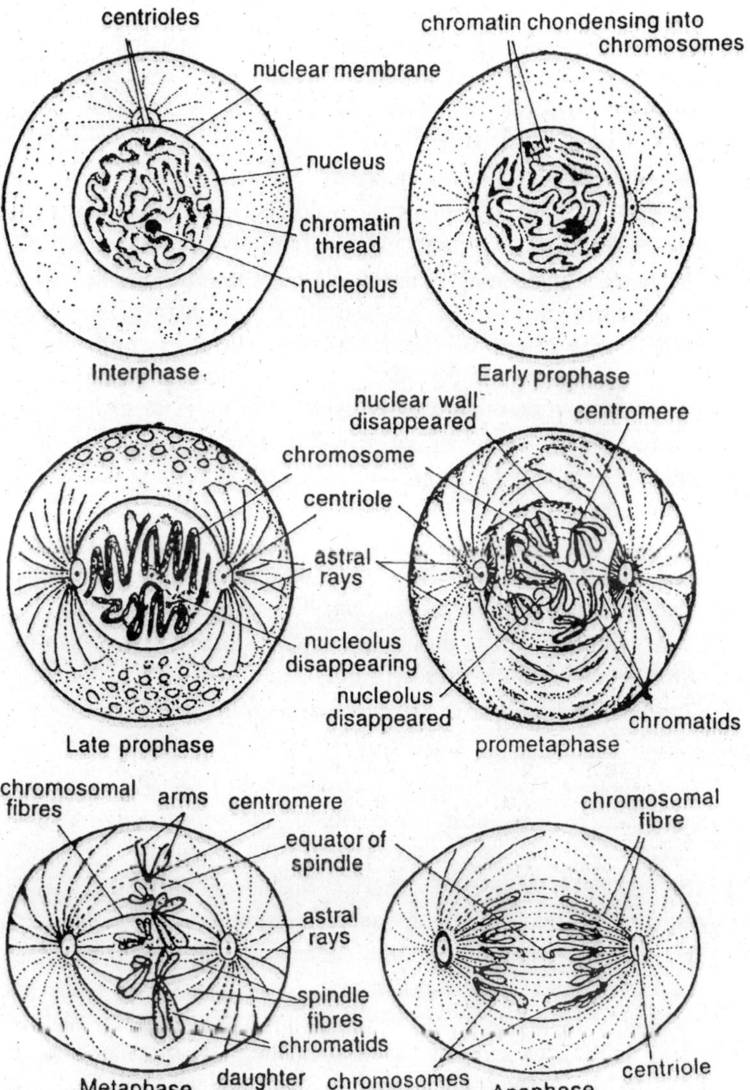

Fig. 12.3. Mitotic cell division in animal cells.

During G_2-phase, all the metabolic activities are performed. M-phase is the period of chromosomal division.

The relative lengths of these phases differ in different organisms. A human cell in culture at 37°C, completes the mitotic cycle in about 20 hours and the M-phase lasts for only one hour. Temperature and cell environment plays an important role in determining the rate of cell division. Even the non-meristematic cells can sometimes be made to divide by changing the environmental conditions. Those cells which are not going to divide any more, have the mitotic cycle at the G_1 phase and start differentiating.

The cells shows following changes:

1. The cell, as a whole, attains the maximum growth and possesses synthesized proteins for energy for various divisions and processes.
2. The nuclear membrane is intact and the chromosomes are found in the form of more or less loosely coiled threads, somewhat closely appressed to the membrane. In this condition of chromosomes, most of the cytologists regard them to be duplicated, while some workers are of the opinion that they are multipartite.
3. The two centrioles, which are found at right angles to each other replicate into two each. *Mazia* (1961) has described that, if the replication is checked, then division will not take place.
4. For the future spindle condensation of protoplasm into a coherent area of jelly-like consistency also takes place. The spindle also starts growing and pushes the centrioles apart.
5. DNA synthesis occurs during autosynthetic interphase, when the chromosomes and disperesed.
6. Chromocentre are also conspicuous during the interphase.

Prophase

The prophase is the longest in M-phase and may take about one to several hours. In the neuroblast cells of grasshopper it takes about 102 minutes. Various important changes in this stage are as follows:

1. The cell tends to become spherical and increases its refractivity and turgidity by increasing its surface tension by removal of water.
2. The nucleus imbibes water from cytoplasm, and the chromosomes get themselves organized differently. Each chromosome shows its peculiar structure with marked isolation of chromatids, each of which undergoes a regular cycle of coiling. At the beginning of division the chromosomes start to contract, get thickened and undergo their coiling. Partially this change appears to be associated with the development, at right angles to old gyres of the coils. *Sparrow* (1941) referred that total end to end contraction is approximately one fifth of the initial length. Physiologically, the coils are formed due to continuous condensation. The coils are of two types: smaller minor coils and larger somatic coils. The mode

of twisting during the formation of coils and gyres is also categorized in two distinct varieties:

(i) *Plectonemic*. Twisted in such a way that it is not easy to isolate them, and

(ii) *Paranemic*. Coiled chromatids can be separated laterally.

3. The chromatids are connected with centromere. The chromosomes are separately distributed in the nuclear cavity. The RNA and phospholipid contents gradually increase.
4. The nuclear boundary becomes disrupted, the spindle apparatus begins to form, and the nucleolus and the centromere generally disappear. The spindle formation takes place in two ways viz.,

(i) Single centriole divides into two daughter centrioles and on separation, astral rays appear as delicate filaments, called the spindle. The centrioles migrate along with asters until they become located at antipodal positions. This type of spindle is referred as to the *central spindle*.

(ii) The two centrioles are already polarized before the beginning of the division and the spindle formation takes place at metaphase. This type of spindle is known as the *metaphasic spindle*.

The mitosis in which achromatic figure and spindle are formed by centers is called as the *amphiastral mitosis* and whence centers are absent, the mitosis is called as '*anastral*'. The anastral mitosis takes place in plants.

Prometaphase

This stage follows the complete disappearance of the nuclear membrane, the chromosomes tend to aggregate in a central position in the cell, near the equator. The term prometaphase is described by *Coin* (1964). *White* (1963) defined it as the period during which the spindle is being formed and during which the chromosomes give the impression of struggling and jostling one another in an attempt to reach the equator of developing spindle. At least, in plants, this stage corresponds to the first appearance of an organised spindle. *Wilson* and *Hyppio* (1955) considered this positioning of chromosome to play an important role, both in the development and functional organization. The spindle fibers are tubular, elastic, fibrous and proteinous in nature.

Metaphase

The metaphase chromosomes are sharply defined and descrete bodies and are tightly coiled. In this stage the chromosome number can be

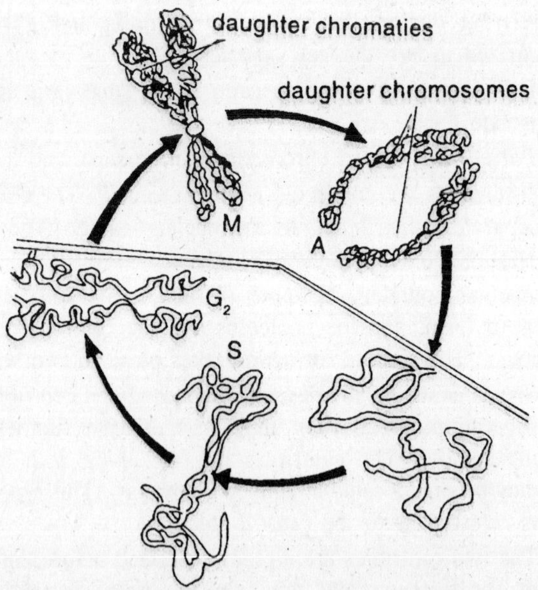

Fig. 12.4. Changes in chromosomes during cell cycle.

easily counted and it is possible to recognised the different chromosomes from their sizes, shape and gross structure. In early metaphase characteristic spindle-shaped figure appears in the clear nuclear region. This consists of fibrous radiations which extend from the central broad-portion called the *equator* and converge at two points on the opposite ends of the nuclear region, called the *poles*. The chromosomes which were till now scattered in a random manner in the central nuclear portion of the cell begin to show peculiar movements and arrange themselves in a single plane midway between the two poles of the spindle, forming an equatorial plate. To be more precise it is the centromeres of the chromosomes that are lined up at the equatorial plate. Usually, the smaller chromosome occur near the center of the *equatorial plate* and the larger ones near the outer ends. It is, however, not necessary that the two homologous should be near each other on the equatorial plate. The position of each chromosome is independent of other on the equatorial plate. The metaphase chromosome is a visibly double structure and is clearly seen to be split longitidinally into two exactly identical chromatids.

The centromere plays an essential role in the development of the spindle and the separation of the daughter chromosomes. The chromosomes are so arranged at the equator that one chromatid of each chromosome faces one pole and the other faces the opposite pole. Two spindle fibres are attached to the centromere of each chromosome, one on either side of it. These connect the chromosome to the two opposite poles of the spindle and are called *chromosomal* or *tactile fibres*. Other fibres of the spindle extend from pole to pole and are not attached to the chromosomes. These are called the *continuous spindle fibres*. The spindle fibres are composed mainly of protein, some ribonucleic acid (RNA) and lipids.

Form and arrangement of the chromosomes

The arrangement of chromosomes at the equator of spindle is not of one type in all organisms, as the arrangement depends upon the shape, size and number of chromosomes that differ in different organisms, because in some organisms they (chromosomes) are thread like (Urodela and in insects (Orthoptera and Diptera)). In Odonata, Coleoptera and Hemiptera they are shorter or rod shaped while in Orthropod they are rounded.

On the basis of their morphological study, one can divide them into three categories:

(i) *Straight rods or threads*. These chromosomes arise directly by shortening of the spireme thread.

(ii) *Loops, V's or hook forms*. These chromosomes are formed by a flexture at the middle point or near one end.

(iii) *Ovoidal or Spheriodal forms*. These forms arise by the extreme shortening of threads.

All the three forms mentioned above are double at metaphase due to their longitudinal splitting. The longitudinal splitting among rod shaped and thread shaped chromosomes, can be traced clearly but in spheroidal forms, it often appears as an apparently transverse constriction due to the extreme shortening of chromosomes.

Spindle attachment of the chromosomes

The mode of the attachment of the chromosomes to the spindle fibres based on the structure of chromosomes. But even though, the arrangement and the way of attachment is constant for every particular chromosome and that it is inherited from generation to generation. Chromosomes are attached to spindle fibres by centromeres. The chromosomes lacking centromere fail to attach with spindle fibres. The way of their attachment may be of two types:

(i) Terminal or telocentric

(ii) Non-terminal or atelocentric.

In terminal attachment the chromosomes may have their connection with spindle fibres at the free end. Non-terminal attachment may be at the middle point (*median*) or at an intermediate point *sub-median* or *sub-terminal*.

Anaphase

In anaphase the pairs of centromeres move apart along the spindle and carry one daughter chromosome of each pair to opposite poles. The spindle eventually grows longer. The movement of chro-matids is complex. First separation of chromatid from centromere takes place and then a flow of current along the spindle governs their poleward movement. This stage lasts for very short duration, varying between 6 to 12 minutes. In the last anaphase the zone between the two sets of chromosomes or the equatorial region gradually increases. In seems that the fibres stretch and are named *interzonal fibres*. The expansion of middle part has been referred as the stemmkoper or pushing body by *Bear*. The pushing body looks like a gel which pushes the chromosomal sets towards respective poles. During the poleward movements the chromosomes assume peculiar 'J' or 'V' shapes, depending upon the position of centromere. At this stage, 'J' are called *heterobranchials* and 'V' chromosomes as *isobranchalis*.

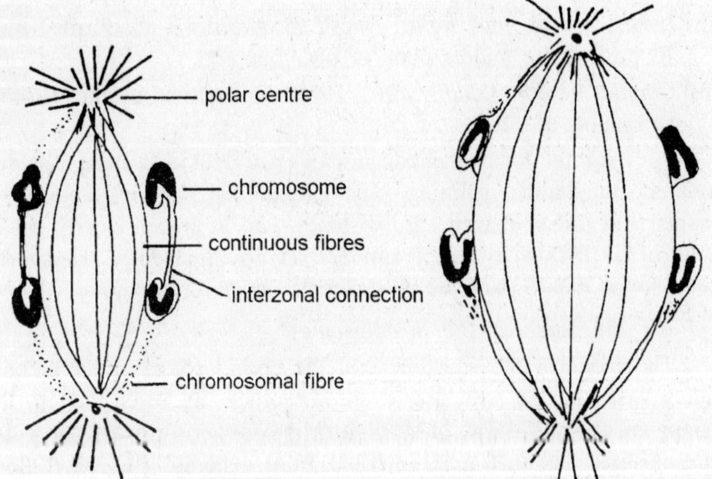

Fig. 12.5. Two types of anaphase spindles. A-direct types, B-indirect type.

Movement of chromosomes

The movement of the chromosome are controlled by spindle fibres. Actually two processes may be at work; a continuous enlargement and elongation of the spindle and the shortening of the chromosomal fibres. Since the fibres shorten without becoming thicker, the process probably involves the removal of water or other molecules from fibres. A band of 'interzonal fibres' is often seen for a time after the separation has been accomplished, connecting the chromosomes which have pulled apart, and often including a remnant of the spindle.

Forces involved in the movement of chromosomes

Spingle fibres are responsible for the movement of chromosomes during anaphase i.e., from equator to the pole. Several models have been porposed by diferent workers to explain about the forces involved in the movement. Some of the them are given below:

1. *Simple contraction model. Van Benden* (1883) suggested that chromosomes in the dividing cell are pulled towards the poles by the contraction of spindle fibres. *Swann* (1962) described that centromere secrete some substance which causes the contraction of spindle fibres. The main objection to the theory comes from the direct observations on chromosomes being carried even beyond the centriole to which the spindle fibres are attached. Further during cell division the whole cell elongates which is just opposite to the contraction model.
2. *Expansion Model. Watase* (1981) suggested that the spindle fibres exert pressure on the nucleus and chromosome become flattened on the metaphase plate. The chromosomal fibres become attached to the chromosomes, which are now pushed towards the opposite poles. For this reason, this is also known as *pushing model.*
3. *Contraction and expansion model. Belar* proposed that initial separation of daughter chromosomes is an autonomous process but further separation is the result of both contraction or expansion of various parts of spindle. The chromosomal fibres which extend from the centromeres of chromosomes to the poles of spindle contract and pull the attached chromosomes to the poles. The interzonal fibres present between the separating daughter chromosomes expand and push the daughter chromosomes towards opposite poles.
4. *Equilibrium dynamic model.* This provides the most convincing explanation about the possible mechanism of chromosome movement. *Inove* and *Sato* (1967) have described the occurrence of an equilibrium between a large pool of monomeres which constitute the proteins of microtubules. During polymerization some hydrophobic interactions occur between the non-polar groups of

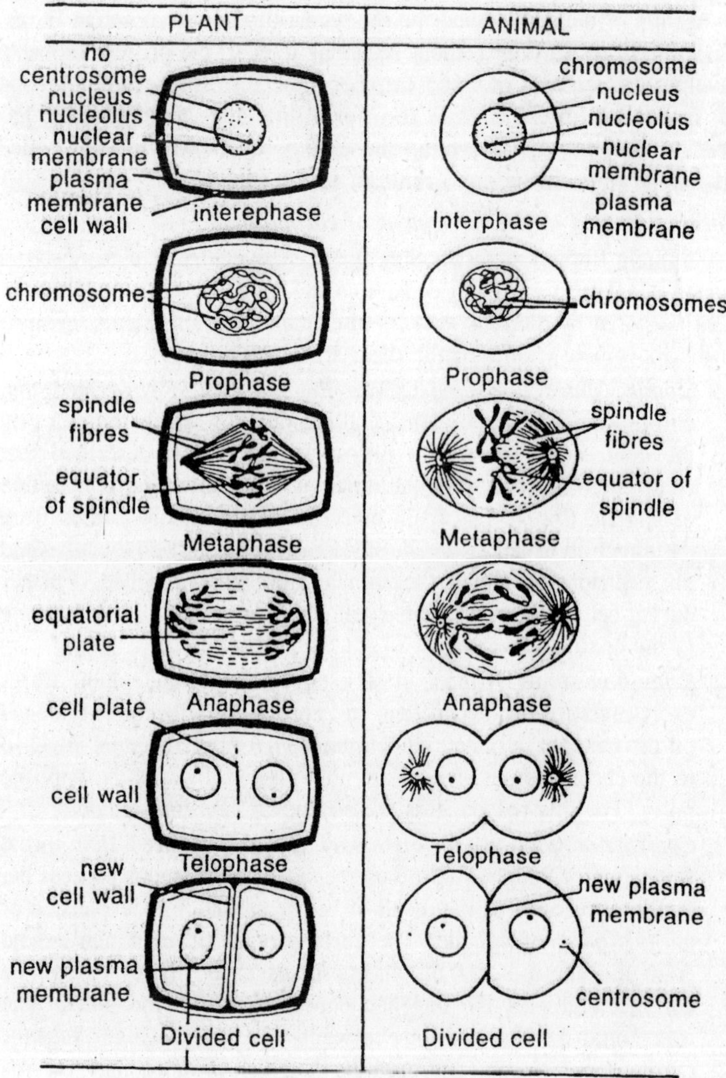

Fig. 12.6. Differences in mitosis in plants and animal cells.

protein monomeres. During chromosomal movement both contraction and elongation of spindle fibres occurs by the substraction or addition of few monomeres to the fibres. During anaphase chromosomal fibres contract by the delation of monomere from the polar end of the fibres and the continuous fibres increase in size by the assembly of new material at the polar ends. Thus the spindle elongates and distance between the poles increases.

5. *Gliding model. Ostergen* and other are of the opinion that the movement of chromosomes is an active process in which chromosomal fibres glide as sail boats between the continuous fibres. *Ambrose* has suggested that motile forces for these movements could be electrosmosis and electrohoresis. Break-down of ATP is supposed to provide the magnitude required for spindle movement.
6. *Ratchet model. McIntosh, Helper, Van Wie* and others have described the presence of mechanical bridges between the microtubules similar to the ratchet connections in the muslces fibres. The chromosome fibres glide between the continuous fibres and draw the chromosomes apart. The energy is supplied by the break-down of ATP.
7. *Electrical model. Lillie* and *Coller* (1936) suggested that the change in the membrane potential is due to local changes in permeability occurs near the poles and around the nuclear membrane and produces electric field. The chromosomes are negatively charged in prophase and the charged chromosomes easily migrate in the electric field.
8. *Push Model.* According to this concept chromosomes are pushed apart by the colloidal component of cell cytoplasm. The colloid imbibes water, swells up and pushes the two chromatids of the chromosomes apart. The spindle fibres act as tracks that guide the movement of chromosomes to the poles, preventing their scattering throughout the cell.

Telophase

The two groups of chromosomes aggregate at the pole and on arrival they progressively lose their chromaticity. Decondensation takes place. Reformation of the nuclear membrane by certain unknown process take place. Perhaps new material is synthesized by RNA during telophase or it may be possible that the endoplasmic reticular system may be giving rise to a new membrane round the nucleus. All chromosomes become uncoiled. The spindle fibers are absorbed in the cytoplasm. The nucleolar organizer or SAT zone again forms the nucleolus. Finally, these change result into two daughter nuclei which are equivalent to the parent nucleus in all respects.

Cytokinesis in plants

Nuclear division or mitosis, as it is called it followed by the division of the cytoplasm. While the daughter nuclei are being organised at the poles, the mitotic spindle disappear except at the equator, where the continuous spindle fibres become more dense. This region is now called the *phragmoplast*. According to *Porter* and *Machade* (1960) the formation of cell plate is initiated by migration of tubular elements of the endoplasmic reticulum toward the interzonal region of the spindle where they spread out to form a close lattice along the mid line.

Theories of cytokinesis *Singh & Tomar* appears in phragmoplast and contain pectic substances which fuse to form in the middle of dividing cell a partition known as *cell plate*. More droplets are added to the cell plate to form the middle lamella which now begins to extend outwards until it reaches the outer wall of the original cell. Primary cell wall material is now deposited on either side of the middle lamella. This divides the cytoplasm into two roughly equal parts. Cell division is now complete and two cells are formed. The division of the cytoplasm is called cytokinesis.

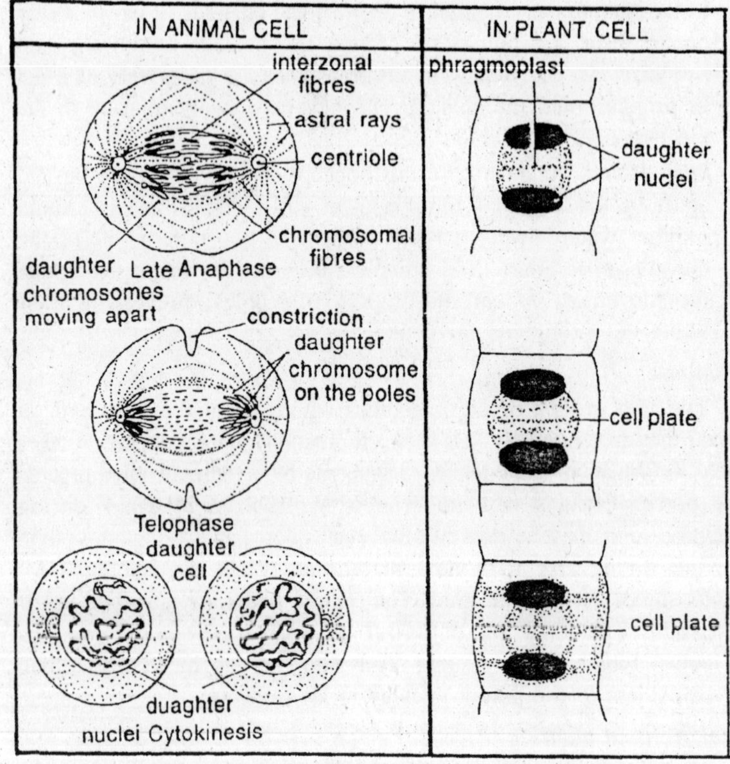

Fig. 12.7. Division of cell cytoplasm in animal and plant cell.

Cytokinesis

Cytokinesis in Animal Cells

Cytokinesis starts by the appearance of a shallow furrow in the cytoplasm at the equator of the spindle. Slowly and slowly the furrow deepens and constricts the cytoplasm and the cell into two daughters. A number of theories have been proposed for the formation of furrow. These are as follows:

1. *Contractile Ring Theory*. According to *Swann* and *Mitchison* (1958) the cytoplasm around the equator of the spindle contains some contractile proteins. These proteins form a sort of ring at the equator. As the dividing cell elongates, the contractile ring contracts which results into the furrow formation.

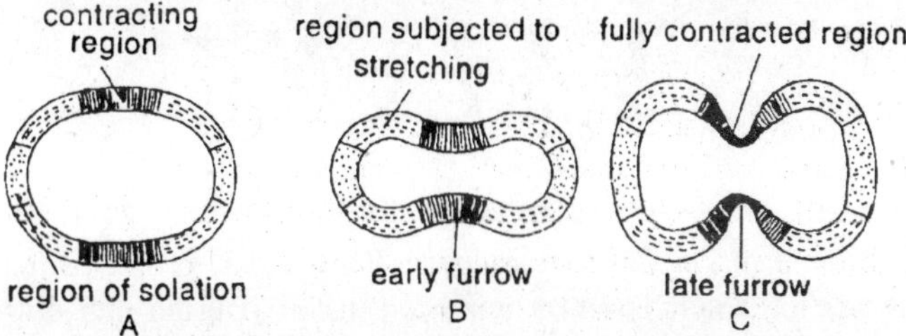

Fig. 12.8. Diagram to illustrate the contractile ring theory of Marsland for cytokinesis. A–Early telophase, B–Early furrows, C–Late furrow.

2. *Expanding surface theory*. *Mitchison* (1922) proposed that a nuclear material is liberated by chromosomes which is responsible for the cellular expansion at the poles. As the polar areas expand, the equator contracts which results in the appearance of furrow. The furrow divides the cell into two daughter cells. There are certain examples, showing that the furrow formation takes place in the absence of nuclei or chromosomes (*Nachtway*, 1965). This indicates that exciting material might have originated from other than nucleus.

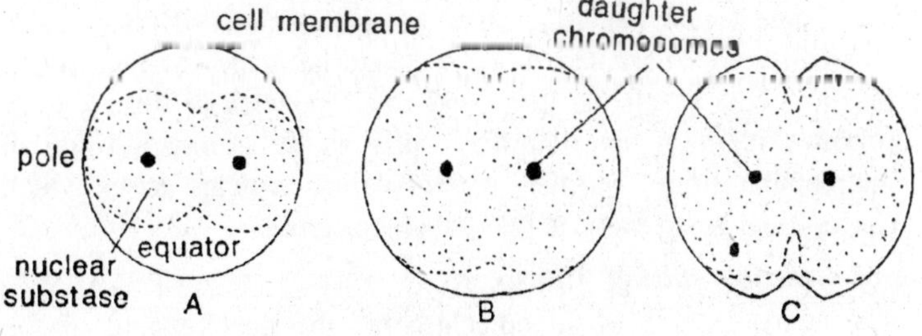

Fig. 12.9. Diagram to illustrate the expanding surface theory.

3. *Spindle elongation theory.* This theory was proposed by *Dan* and *Dan* (1947) and *Dan* (1958). According to them the spindle and asters are responsible for the cytokinesis. During experiment, kaolin particles were attached with the egg membrane. It was observed that elongation of the cell at anaphase is accompanied by shrinkage of equatorial plane results two kaolin particles on either side of equator come close to each other. The driving force is believed to be the elongation of microtubules of the spindle which push the centres apart.

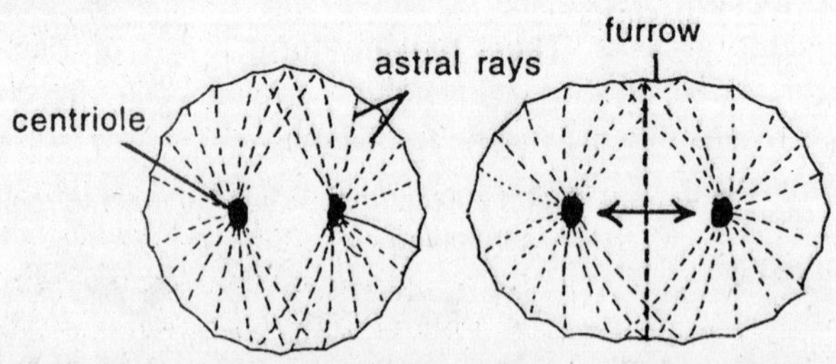

Fig. 12.10. Diagrammatic representation of Dan's spindle elongation theory.

4. *Astral relaxation Theory. Wolpert* (1960, 1963) proposed that the surface of a cell is under uniform tension. During cell division, as the astral rays reach the poles they lower in the surface tension at the poles. The surface tension at the equator remains same. Because of low surface tension, the polar regions expand and cause the appearance of furrow at the equator.
5. *Vesicle Formation Theory.* According to *Threadgold* (1968), during anaphase, dividing cell lengthens. This occurs in inter-zonal region due to increased electron density. Plasma membrane also exhibits high electron density in the furrow simultaneously, the continuous spindle fibres, present in the inter zonal region, continue to increase its density and finally form an adielectronic fibrillary plate in the equator. As furrow progresses, a large, empty membrane-bound vesicle appears on each side of this plate. Later on small vesicles are added in the equatorial plane. At final stage, fusion occurs of all the large and small vesicles ot form a deep furrow, leaving the daughter cells to be connected by an intercellular bride. Further, a row of tiny vesicles arises which causes final separation of two daughter cells.

Role of Centriole during Mitosis

The centriole acts as an epicentre for the development of the central spindle. Whether the centriole produces the spindle fibres or serves to direct their formation or on the other hand is entirely passive in the process, cannot be uniquestionable deduced from all the informations at hand. According to *Cleveland* (1957) the spindle fibres and astral and chromosomal fibres are recognizable first in the close vicinity of the centrioles. The completely developed achromatic apparatus then functions to separate daughter chromatids towards respective poles. With relation to the mitosis, *Cleveland* concluded two important points: (i) relative temporary independence of the achromatic and chromatic apparatus as regards apparent duplication and function; (ii) final dependence of the cell upon the integrity of these two system and their united function.

Duration of Mitosis

The time required for mitosis differ with species and environment. Temperature and nutrition, in particular, are important factors. The entire sequence of phases may be completed in 6 minutes to many hours. Normally the entire cycle of cell division takes approximately 18 hours; about 17 hours for the interphase. Different phases of mitosis are of different duration. Anaphase is the shortest, the prophase and telophase the most prolonged, and the metaphase of intermediate duration.

Significance of Mitosis

1. *Equal distribution of chromosomes.* The essential feature of mitosis is that chromosomes are distributed equally among the two daughter cells. With every cell division there is a division of chromosomes. The constant number of chromosomes in all the cells of the body is because of mitosis.
2. *Surface-volume ratio.* Mitosis restores the surface-volume ratio of the cell. A small cell has a greater amount of surface available in relation to volume than a large cell. As the cell increases in size the available surface area in relation to increased volume becomes less. By undergoing division the cell becomes smaller in size and the surface-volume ratio is restored.
3. *Nucleoplasmic index.* The growth of a multicellular organism is due to mitosis. A cell cannot grow in size to a large extent without disturbing the ratio between the nucleus and the cytoplasm. After a particular size has been reached the cell divides to restore the nucleoplasmic index. Thus growth takes place by an increase in the number of cells, rather than by increase in the size of the cells.

4. *Repair*. The repair of the body takes place because of addition of cells by mitosis. Dead cells of the upper layer of the epidermis, cells of the lining of the gut, and red blood corpuscles are constantly being replaced. It is estimated that in the human body about 500,000,000,000 cells are lost daily.

Abnormalities in Mitosis

Under unfavourable conditions the mitotic divisions become defective and thus various abnormalities are formed when exposed to physical and chemical agents like temperature, radiations, necrotics and enzyme inhibitors etc. Some abnormal mitotic divisions are given under the following heads:

C-Mitosis (abnormal spindle formation)

Brachet (1975) has described that colchicine inhibits mitosis by disorganizing spindle formation. The results are the formation of polyploid cells, and reduplication of chromosomes. Besides, during mitosis, sometimes cell nucleus is either deformed into a single spheriod compact mass (pycnosis) or becomes broken down (karyorrhexis).

Cytasteral mitosis

Wilson (1901) noticed the formation of numerous asters in the cytoplasm of unfertilized eggs called cytasters by placing them in hypertonic sea water. In the eggs of marine invertebrates, production of many small cytoplasmic asters is of frequent occurrence. The intermixing of nucleoplasm and hyaloplasm seems to be neccesary condition for the aster formation. *Costello* (1940) showed that asters can be produced only if germinal vesicles has broken down.

Multipolar and Catenar mitosis

Mitotic division showing several spindles and centrosomes are common in many protozoans and fish eggs. Multipolarity is usually caused by uneven division of centrosomes as well as irregular distribution of chromatids of the different spindles. It results in the formation of cells as aneuploidy (uneven chromosome number). Catenar mitosis have been described by *Dalcq* and *Simon* (1932) in amphibian eggs. During this, dividing cell forms a large number of asters are capable of dividing in the absence of spindle or nucleus in an autonomous way.

Achrosomal mitosis

Sometimes, mitosis in induced by various agents without use of nucleus or chromosomes. *Briggs et. al.* (1951), in fertilized eggs with heavily X-rayed sperms, removed the egg's maturation spindle by pricking and sucking. In this way, he obaïned very nice non-nucleate blastulae. Similarly, *Stauffer* (1945) obtained normal nonnucleated Axolotl blastula.

Anastra mitosis

In plant cells and many oocytes, generally asters around the centromeres are absent, thus producing anastral mitosis. *Bataillon and Tchou Su* (1933) described in details the anastral mitosis in amphibian interspecific hybrids.

Achromatic Figure

Achromatic figure consists of spindle fibers and astral rays. Astral rays are developed just before the distintegration of nuclear envelope during late prophase, while the formation of spindle rays can be traced only after the disappearance of nuclear membrane.

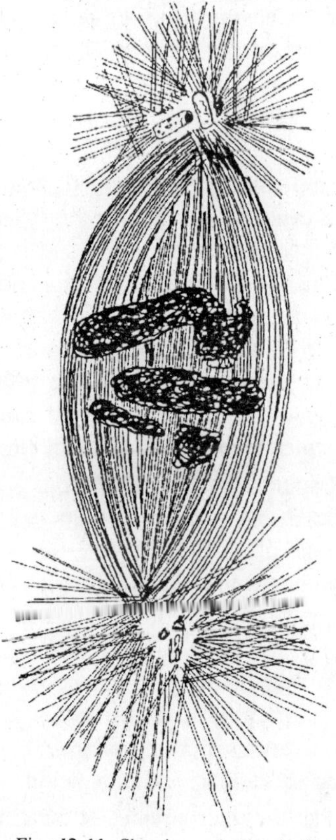

Fig. 12.11. Showing achromatic figure.

The spindle

In fixed preparations of dividing cells, fibers are seen forming a spindle shaped body lying between the centrioles. It is generally believed that spindle fibers are not visible in normal living cells. They can be produced by adding acid to the medium, and, on removing the acid, the fibres disappear; as most fixatives contain acid, it is stated that the fibers are artificats. There has been considerable controversy as to the reality of fibers and various theories have been put forward to explain their presence in fixed material. *Schrader* in 1944 observed the spindle fibers in living cleavage cells of mite and certain flagellates.

The spindle is superficially similar to the spindle shaped area found between the poles in the magnetic field. The spindle consists of two kinds of fibers which have been supposed to differ widely in functional significance.

Those spindle fibers extending from the poles to the chromosomes are called half spindle, incomplete fibers or discontinuous fibers or chromosomal fibers. The remaining spindle fibers extend without interruption from pole to pole are called the continuous fibers or direct or complete fibers.

It is believed that the spindle is usually formed in two parts. A part arises between the centrosomes before the disappearance of the nuclear membrane; it is cytoplasmic in origin and is called the central spindle. As the nuclear membrane disappears the central spindle moves into nuclear region. Then the nuclear material undergoes gelation around the central spindle to from the 1/2 spindle fibers or chromosomal fibers. In the metaphase the chromosomes are attached to the 1/2 spindle fibers at the equator. *Schrader* suggested that the spindle may consist of protein molecules arranged in parallel lines.

It appears that the spindle is organized from a pre-existing pool of protein. It is postulated that protein monomers aggregate to form microtubules of the spindle fibres. The spindle fibres are thus oriented polymers. It has been proposed that the continuous fibres are elongated by the insertion of subunits within the fibre. This leads to the elongation of the spindle. Conversely, the shortening of chromosomal fibres takes place in the region of the spindle poles. Thus the fibres do not contract like a rubber band, but shorten by loss of material at one end. This is in agreement with observations that no change in the thickness of fibres takes place during shortening and elongation.

There is evidence that the microtubules of the chromosomal fibres originate at the kinetochore. Micotubules are also found in cells which have no centrioles e.g., those of higher plants. Here the microtubules end in the pole region and are not attached anyhwere.

The asters

Astral rays are developed by the centrosome present in the viscinity of nuclear region. Astral rays together with the spindle form the *amphiaster*. In early stages the astral rays are straight, simple and few in number. But in later stages they rapidly elongate in all directions, and increase in number. In certain cases the astral rays of both asters cross one another in the equatorial region outside the spindle. Sooner or later, the crossing of the rays disappear and rays adjust themselves in such a way that their free ends move towards the opposite asters and meet in the equatorial plate thus giving the impression of continuous fibres. These continuous fibres which are formed by the fusion of astral rays are lying in the region outside the spindle. It has been suggested that the long protein molecules of the protoplasm become arranged in rows parallel to the lines of protoplasmic flow and thus constitute the astral rays.

13

MEIOSIS

Mitosis is a device of increasing the number of cells. During asexual reproduction it is the means of multiplying the individual. Therefore, a sexually reproducing organisms, set of characters are transferred unchanged from parents to offspring, except of course accidents. There is uniparental indheritance due to mitosis. But in sexually reproducing organisms, gametes are produced as a result of meiosis. Two gametes unite to give rise to a zygote which divides mitotically to form the individual. More often than gametes from different parents unite to form a zygote, thus ensuring biparental inheritance. This result into a blending of characters from two parents into their offsprings and ensures variations which is important for evolution. Gametes contain half the number of chromosomes, as compared to somatic cells. Union of two gametes restores the somatic number of chromosomes in the zygote. Gametic union results into the mixing of cytoplasms of two cells which is followed by the fusion of two nuclei. During meiosis or at the time of gamete formation, therefore, halving of the number of chromosomes or the amount of DNA is achieved by two successive divisions of cytoplasm and nuclei (meiosis I and meiosis II) accompanied by onle one replication of chromosomes. Somatic cells are derived from the zygote due to mitotic divisions. Zygote is the result of a fusion of two gametes, each of which contain only 23 chromosomes. Both the male and female gametes contain the same amount of DNA and equal number of morphologically similar chromosomes. The zygote and somatic cells derived therefrom, therefore, contain 23 pairs of homologous chromosomes, one of each pair derived from the male parent through the male gamete and the other from the female through the female gamete. The members of each pair are homologous to each other and non-homologous to the other chromosomes. In some organisms, the chromosome complements of male and female cells are slightly different and this forms the basis of sex differentiation. Those cells which possess two sets of chromosomes, are called diploid and they have 2n or dipoid or unreduced or zygotic number of chromosomes. Gametes, on the other have n, or haploid or reduced or gametic number of chromosomes. In sexual reproducing organisms the diploid and haploid phases of the life cycle alternate and this phenomenon is called alternation of generation. In plants, there is an alternation of sporophytic (2n) and gametophytic (n) generatin. Reduction division heralds the beginning of gametophytic generation and fusion of gametes that of gamete formation which occurs in specialised organs i.e., the embryo sac and the pollen sac. A single cell undergoes meiosis in the embryo but the pollen sac a large number of spore mother cells divide meiotically, almost synchronously, to produce hapoid pollen grains. It is, therefore, easier to study meiosis is antheres, similarly testes in animals.

Meiosis consists of two divisions which take place one after the other, during which the number of chromosomes is halved. The two divisions are known as the first meiotic division or heterotypic division and the second meiotic division or homotypic division. The first meiotic division has the smae stages as mitosis. These stages are known as prophase I, prometaphae I, metaphase I, anaphase I and telophse I. Except for prophase, the other stages are similar to those of mitosis. Following telophase I is period of interphase or rest. The second meiotic division consists of prophase II, metaphase II, anaphase II and telophase II.

Heterotypic or 1st Meiotic Division

This is the first half of meiosis and is true reductional in nature. Here the diploid cell divides into two haploid daughter cells. As already stated that the phases of this division one called prophase I, metaphase I, anaphase I and telophase I.

Prophase I

It is the longest stage and its duration makes from a fe hours to a few days. The nucleus starts increasing in volume and this increase is much than the mitotic increase to some extend. This is due to an increase in hydration.

Nobel and Ruttle (1936) have called this stage as the *premiotic spiral prophase*. Prophase I is Further divided into six substages.

(i) Pre-leptotene (iv) Pachytene
(ii) Leptoten (v) Diplotene
(iii) Zygotene (vi) Diakinesis

Preleptotene

As noted previously, replication of DNA has altready taken place during interphase. The chromosomes at the beginning of meiosis are, therefore, note single but double strands. The beginning of meiotic prophase is marked by an increase in nuclear volume. During preleptotene the chromosomes are very thin and cannot be easily seen, except sometimes for the sex chromosomes.

Leptotene ("Slender thread") or Leptonema

During this stage the chromosomes become more distinct, and their double nature is seen in many organisms, Leptotene extends from the time single unpaired chromosomes are first seen to the time when they begin to undergo homologous pairing. Under the light microscope the chromosomes appear as slender threads bearing a series of granul-like structures called chromomeres. Whole mount electron microcope preparations show that leptotene chromosomes have an axial filament to which chromatin fibres are attached as a series of lateral loops. In certain regions these loops appear in bunches, and from the chromomeres of light microscopy. the axial filaments are apparently composed of two independent protein fibrils, one for each chromatid. They later become transformed into the lateral elements of the synaptonemal complex. In the leptotene cells of the quail there is evidence of premeiotic pairing. Two homoloogous chromosomes become apposed to each other, chromomere for chromomere, even though no

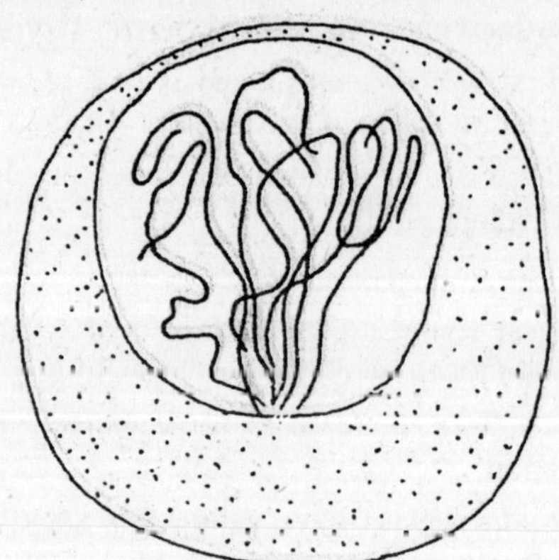

Fig. 13.1. Leptotene chromosome arranged in a 'bouquet'.

synaptonemal complex is formed. In leptotene cells there is a tendency for the ends of homologous chromosomes to be associated with each other at the nuclear membrane, where they are attached.

Leptotene chromosomes may be irregularly arranged, or may be polarized toward the centrioles forming a 'bouquet.' Electron microscopic studies have shown that bouquet formation results when a group of chromosomes is attached close together on the nuclear membrane.

During leptotene the cytoplasm has many polyribosomes, but endoplasmic vesicles are few. In maize (*Zea mays*) the nucleus increases in size during leptotene, reaching its maximum size during pachytene. This increase in size is correlated with synthesis of RNA and protein.

Zygotene ("mating thread") or Zygonema

The chromosomes becomes much shorter and thick and number of coils is also increased. Homologous chromosomes begin to pair by force of attraction. In synapsis or pairing, two homologous chromosomes come close to each othe rprecisely chromere to chromere. The pairing starts from one point and proceeds onward in a zipper like fashion. Out of the two synaptic chromosomes, one is of paternal origin while the other is maternal. These are called *bivalents*. Since each chromosome is made of two chromatids, each pair of chromosomes consists of four chromatids thus representing '*tetrad*' of chromatids at this stage.

The pairing of zygonemal stage is completed in three different ways, as follows:

(i) *Proterminal synapsis*. The two homologous chromosomes start pairing at the terminals, which gradually then progress towards the centromere.

(ii) *Procentric synapsis*. The pairing starts near the centromere and progresses onwards on either side.

(iii) *Random synapsis*. The pairing may be random at many points which then finally fuse together.

Two views have been advocated to explain the mode of initiation of synapsis. *Darlington* (1937) proposed precocity theory which explains that the chromosomes undergoes synapsis due to singleness, but this theory does not explain the extra synthesis of DNA and duplication at leptotene stage. Anothe rview is proposed by *Sax* (1932) which is referred as the retradation theory. It explains that pairing takes place due to the retardation or cessation of metabolic activities of the cell.

During zygonema, the nucleus remains intact and the centriole moves apart initiating the aster formation.

Pachytene "thick thread" or pachynema

It generally begins when the pairing of chromosomes ends. It represents one of the longer substages of prophase I and characterised by the following important events:

When the pairing has completed, the apposition of homologues becomes so tight that it is difficult to identify two separate chromosomes. Within the major coils, at right angles to them smaller minor coils appear. At the height of pachytene, the pairing chromosomes twist about one another as relational coils and each soon longitudinally splits if no splitting had occured. This coiling is further complicated by a coiling of two chromatids of each chromosome around each other. This coiling put the chromosomes under considerable strain. This means that at this stage, each homologue are called *sister chromatids*. Formerly this element of four chromatids was called a *tetrad* – a name which has now been replaced by *bivalent*. When the chromosomes have longitudinally splitted, transverse breaks occur at the same level in two of the adjacene homologous chromatids. Soon both segments of the chromatids interchange and fuse together with those of homologoues. In this process which is called crossing over, portions of the two chromatids with their genes are interchanged and two remain intact. The new chromatids thus interchanged will be mixed.

After this interchange, the parts begin to separate, repalling each other. However, this separation is not at all complete since the homologous chromosomes remain joined at their point of interchange. This point, where the two are joined, is called *chiasma* (pl. chismata). This chiasma is an immediate consequence of crossing overr and are found in all plants and animal with few exceptions (e.g., males and females silkworm). The number of chiasmata is variable but for one bivalent atlease one chiasma is found. Bivalents with 2,3 and 4 chiasmata are rare. The highest number of chiasmata has been observed in the long chromosomeof the broad bean, *Vicia faba*, where there are probably 12 chiasmata in a bivalent. The average number of chiasmata is known as chiasma *frequency*. This number seems to depend upon the length of the chromosome. Short acrocentric chromosomes have a greater number of chiasmata thatn metacentric chromosomes. In small chromosomes, chiasmata are localized at the distal ends. Nucleus remains as such, being somewhat larger than the proceeding substage.

Cytomixis

It is botanical term which refers to the passage of chromatin from one microsporocyte (pollen mother cell) to another during early prophase. This phenomenon is so widespread in the normal species of hybrids that one wonders if it is a general phenomenon.

Diplotene or Diplonema

The separation of homologous chromosomes initiates diplotene. The synaptic forces of attraction between them lapse due to breakage at one or more point, so that the homologous chromosomes uncoil and start separating. But the separation is incomplete since the homologues are in, contact at one more where the crossing over has already taken place. These points of contact are points known as chiasmata (sing-chiasma; meaning cross) which present crosshaped appearance. The chiasma is the morphological equivalent of genetic crossing over. Their number and position varies in the same pair of chromosomes and in different cells of the same individual.

By the end of diplotene, the chiasmata begin to move along the length of chromosomes from centromere towards the end. This displacement of chiasmata is termed as terminalization. When the termimalization of chiasmata is complete the homologous chromosomes are kept in contact by terminal chiasmata. The degree of terminalization is generally expressed as a coefficient of terminalization (T).

$$T = \frac{\text{Number of terminal chiasmata}}{\text{Total number of chiasmata}}$$

The average number of chiasmata in bivalents is known as frequency of chiasmata (fq).

$$\text{Frequency (fq)} = \frac{\text{Total number of chiasmata}}{\text{Total number of bivalents}}$$

Accodring to *Darlingtion* two types of repelling forces operate on the chromosomes at diplotene. One of the forces is electroegatively charged and operates on the surface of the chromosome throughout its length and the other with electropositive charge is localized on the centromere. The former controls the repulsion of the chromosomes and the later causes distal movement of the chiasmata.

Diakinesis

At this stage the chromosomes are best seen as very short and thick bivalents. The contraction is nearly maximum and the chromosomal pairs are well spread throughout the cell as thought by mutual repulsion. The chromatids of each chromosome are very close to each other and their identity is not distinguishable. Terminalization may be completed in the substage. The bivalents migrate to the periphery of the nucleus. The nucleus generally disappears during this stage but may persist, in reduced size until anaphase. The nuclear membrane disrupts and eventually disappears thus all the chromosomes are set free in the cytoplasm. The spindle fibres are organized with an establishment of the pole.

Prometphase I

Spiralization reaches its virtual maximum in this stage; the nuclear membrane disappears, and the chromosomes become arranged at the equator of the cell to begin the metaphase. Nuclear membrane disappears and spindle and astral rays are formed during this stage.

Metaphase I

The bivalent chromosomes arranged themselves in the equatorial plane so that the two members of each homologous pair are found with their centromeres directed toward opposite poles. whereas in mitosis there is a single centromere which unites the chromosomes, in meiosis there are two chromosomes. Each homologous chromosome has a centromere to which the daughter chromatids are united. The repulsion of the centromere is accentuated, and the chromosome is found on the verge of dividing.

Anaphase I

In this the chromosomes are pulled apart by the movement of their centromeres which remain undivided, with the result that each tetrad breaks up into two component pairs of chromatids or dyads. The dyads separate and move along the tactile spindle fibres to the opposite poles. This is the true reduction division.

Telophase I

The group of chromosomes at each pole forms a nucleus. Each daughter nucleus has half the number of chromosomes as compared to the parent nucleus. Unlike the mitotic telophase each chromosome consists of two distinctly separated chromatids united only at theiri centromeres. This is followed by the formation of a cell plate between the separated sets of chromosomes, dividing the mother cell into two daughter cells. Sometimes, cytokinesis fails to occur after the first meiotic division. In such cases the second meiotic division takes place in the common cytoplasm of the mother cell.

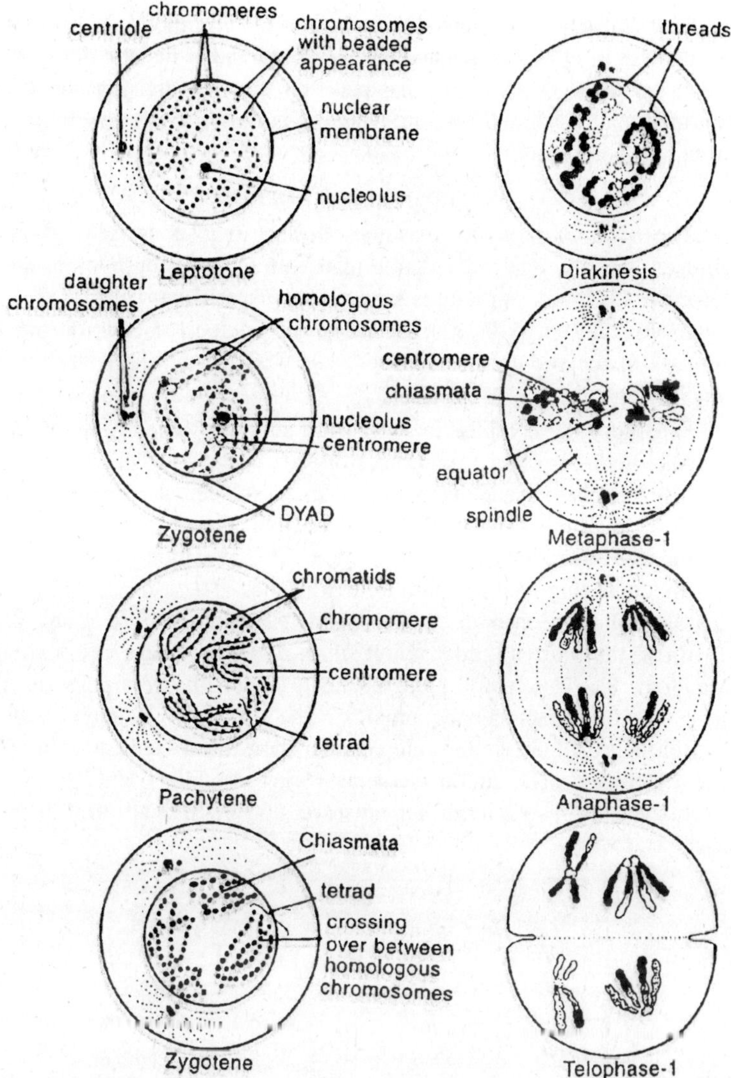

Fig. 13.2. Diagrammatic representation of different stages in the first meiotic division.

Interkinesis

After telophase, two haploid daughter nuclei or cells sometimes undergo typical resting stage as in mitosis. This intervening stage between first telophase and the beginning of second prophase is termed as interphase or interkinesis. This stage is either very short or may be-entirely absent, if present, the coiling of the chromosomes persists. No increase in DNA takes place between the I and II meiotic divisions. The occurrence of an interphase between the two division suggests that additional RNA and protein synthesis is required to carry meiosis through to completion.

Synaptonemal Complex

The behaviour of chromasomes during meiosis is more easily followed with the light microscope than with the electron microscope. However, distinctive structures known as *synaptonemal complexes* are visible with the electron microscope in the nucleus of cells during I prophase of meiotic division. Moses (1950) first of all discovered this complex.

Synaptonemal complex is a tripartite structure found between the two paired homologous chromosomes of each bivalent nuclei undergoing meiosis. Complete synaptonemal complexes are seen at zygotene stage in the region of pairing however, at pachytene they are even more conspicuous under electron microscope. The synaptonemal complex appears consisting of three parallel dense lines. These dense lines are separated by less dense areas. The two lateral elements are composed of fibres that are slightly wider than 10 nm. These are called *Synaptomeres*. The central element has a ladder-like configuration in the centre of synaptonemal complex. The transverse elements have electron dense filaments the interconnect the central element with the lateral elements. The lateral elements varies in width from 20 nm to 30 nm, and the two lateral elements are situated 100 nm to 120 nm apart.

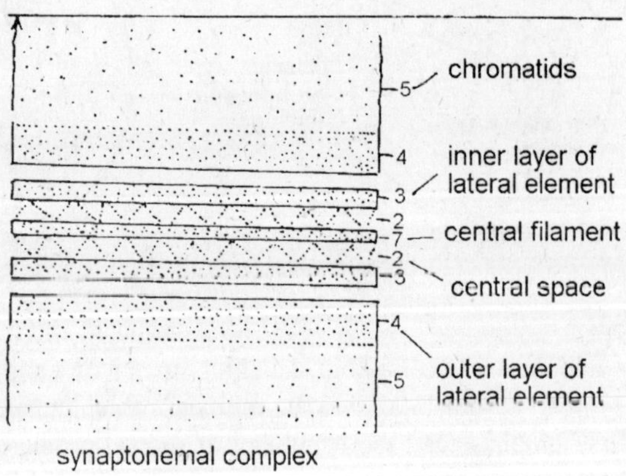

Fig. 13.3. Structure of synaptinemal complex formed during meiotic pairing.

Cytochemical studies have shown that the lateral elements are rich in nucleic acid (DNA, RNA) and proteins, where the central element contains mainly RNA and a small amount of DNA.

According to *Comings* and *Okada* (1971) a series of lateral loops of chromatin arise from the lateral elements. These loops make up the substance of chromosome. Besides another series of smaller loops also arise from the lateral elements. These smaller loops fuse to form the central element of synaptonemal complex.

In 1970, *Robert King* proposed a hypothesis for the formation of synaptonemal complex. This hypothesis is known as, "The synaptomere-Zygosome Hypothesis." According to *King*, there are structures known as synaptomeres, which are coiled polynucleotide segments scattered along the length of a pair of synaposed chromosomes lying in close proximity to one another. Each synaptomere is made out of A, B and C segments. The A and C segments are called *lateral segments* and the B is central. The lateral segments of synaptomeres pair with the respective segments of the adjacent synaptomeres. The B segments are directed towards the central element of synaptonemal complex. These are the sites where zogysomes are attached. The zygosomes are rod-shaped subunits. These are formed in the nucleoplasm and visualized as protein molecules. The zygosomes have a folded head by which they can attach to the B segment of the synaptomere. The tail ends of zygosomes contain charged sites. These charges allow the zygosomes to bind laterally with adjacent zygosomes in a ladder-like manner.

Functions of Synaptonemal Comples

A number of functions have been assigned to the synaptonemal complex. These are :

1. Accoriding to *Moses*, it is prerequisite to chiasme formation and crossing over. But alone it is not sufficient. The presence of synaptonemal complex does not necessarily mean that chainsmata are present.
2. *King* (1970) suggested that synaptonemal complex may orient non-sister chromatids of homologous chromosomes in a manner that facilitates enzymatically induced exchanges between their DNA molecules.
3. Several evidences indicate that it is more directly related to the process of recombination. It has been observed that a small amount of DNA is synthesized at meiotic prophase. If this synthesis is inhibited, the function of synaptonemal complex is arrested.

Homotypic or 2nd Meiotic Division

Before the start of this division, most of the activities of heterotypic division cease for sometime. This resting stage is termed as the interkinesis. 2nd Meiotic division involves following staps:

Prophase II

This stage superfically resembles that the mitosis prophase with following exceptions:

At this early prophase II, the two chromatids of each dyad looks like X, since they are conjoined by a common centromere. The four

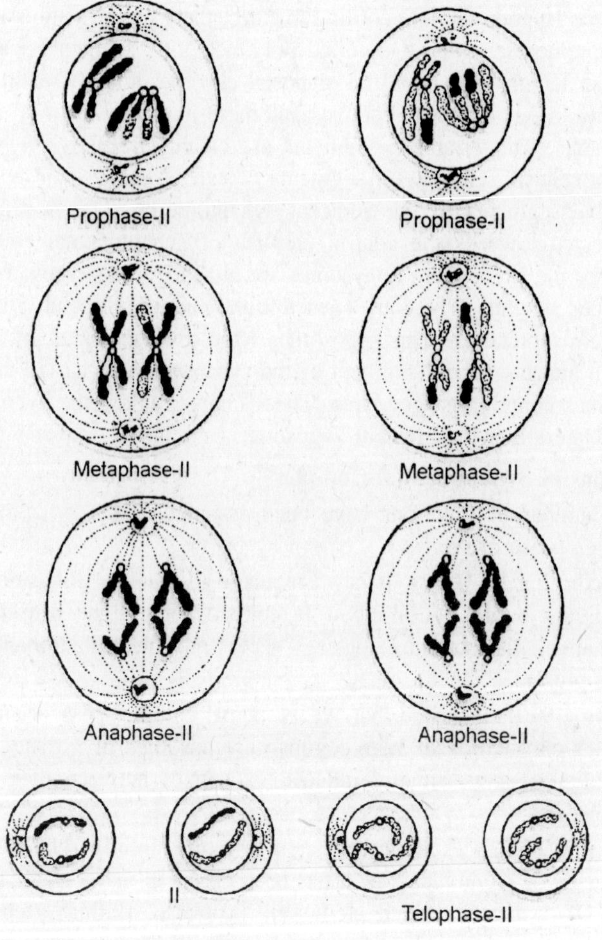

Fig. 13.4. Diagrammatic representation of homotypic division of meiosis.

arms are widely separated. There is no rational coiling. The X-shaped dyad are quite longer than at telophase. The chromonemata are still not completely coiled. The genetic constitution of the two chromatids of each dyad depends upon the kind and number of cross overs which took place in the phase I. If there is no crossing over, the dyads would consist of two identical sister chromatids of either paternal or maternal origin. At the end of the prophase II the nucleolus disappears. The nuclear membrane disappears and the acromatic figure is developed.

Metaphase II

As usual, all the chromosomes arrange themselves on the equator for a short duration. The centromeres touch the equator but their arms radiate out in different directions. Later on the centromere in each dyad divides into two sister centromeres.

Anaphase II

Sister centromeres separate to the poles, pulling with them the chromatids to which they are attached. The chromatids, however, become much thick and stout.

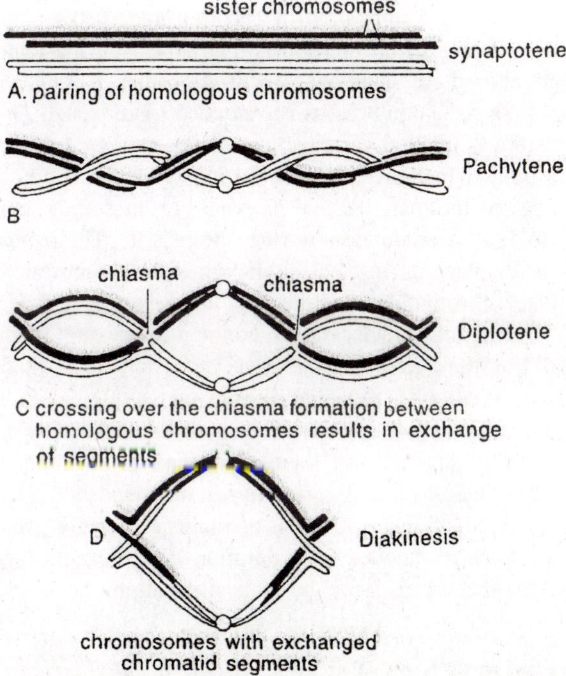

Fig. 13.5. Mechanism of chiasmata formation and crossing over.

Telophase II

The chromosomes at each pole uncoil and thin out to form the nuclea net. Each gorup gets surrounded by a nuclear membrane. Nucleolus reappears. Thus two nuclei are recognised in each cell. This is soon followed by cytokinesis and two cells are formed from each haploid daughter cell. Thus as a result meiosis four cells are produced, each with a haploid set of chromosomes, i.e., each contains just one members of each homologous pair.

Cytokinesis

Sometimes meiosis I is followed by cytokinesis and sometimes it is deferred until the end of meiosis II. The details of the process are same as during mitosis. Cell plate formation results into a tetrad of cells with reduced chromosome numbers. The arrangement of cells in a tetrad is different in different organisms, but is characteristic of the species.

Like mitosis, meiosis too is a dynamic process and its different stages merge into subsequent ones. There is no sharp demarcation between different stages. DNA synthesis takes place during interphase, prior to prophase I. Duplication of chromosomes takes place during interphase itself and the chromosomes that appear during leptotene are double and not single. Number of chromosomes is reduced during meiosis I. If the reduction process is examined more closely, it becomes clear that there is qualitative reduction during meiosis I of those parts of chromosomes which are not involved in exchange. The homologous chromosomes or the maternal and paternal chromosomes segregate or undergo qualitative reduction during meiosis I. Their quantitative reduction takes place during meiosis II, when their centromeres divide and the sister chromatids separate. But the reverse is true of those parts of chromatids which are exchange due to the formation of chiasmata. For these parts quantitative reduction takes place during meiosis I and qualitative reduction during meiosis II. Thus, complete of both qualitative and quantitative reduction takes place as a result of meiosis I and II and none of them alone can bring about complete reduction. It is, therefore, not proper to say that meiosis I is reductional and meiosis II equational. Duration of meiosis varies a great deal in different organisms. So does the duration of different stages in an organism. Prophase I lasts for the longest duration.

Types of Meiosis

There are three types of meiosis, *zygotic* or *initial meiosis*, *gametic* or *terminal meiosis* and *sporic* or *indeterminate meiosis*.

Zygotic or Initial Meiosis

Zygotic or *initial meiosis* (haplotone patter) occurs in lower plants. Fertilization is immediately followed by meiosis, giving rise to cells with the haploid chromosome number. The zygote is the only diploid stage in the life cycle.

Gametic or Terminal Meiosis

Gametic or *terminal meiosis* (diplotonic pattern) occurs in animals. The cells resulting from the two meiotic divisions are transformed into sperms or eggs without further cell division. The process is as follows: Primordial germ cells divide repeatedly to give rise to many generations of daughter cells, spermatogonia (in males) or oogonia (in females). The spermatogonia and oogonia are transformed into primary spermatocytes and primary oocytes, respectively. Each primary spermatocyte undergoes the first maturation (meiotic) division to give rise to two spermatids, which are then transformed into spermatozoa. In the female during the first meiotic division the primary oocyte transformed into a large secondary oocyte and a smaller cell, the first polar body. the secondary oocyte undergoes the second meiotic division and divides into an ootid and a second polar body. The first polar obdy may or may not divide into two. The ootid is the ovum proper.

Sporic or Indeterminate Meiosis

Sporic or *indeterminate meiosis* (diplohaplotonic pattern) is characteristic of higher plants and some thallophytes, but is not known in animals. In the diploid sporophyte, meiosis takes place at some indeterminate point between the zygote and gamete formation. As a result of meiosis haploid megaspores and microspores are formed. The megaspores and microspores undergo repeated mitotic divisions and form the male and female gamet-ophyte, respectively. Certain cells of the gametophyte differentiate and from the gamets. In plats the cells that undergo meiosis are the microspore mother cell and the megaspore mother cell.

Significane of Meiosis

The process of meiosis keeps the chromosome number constant from one generation to other. This also involves reshuffling the hereditary characteristic which arises from reassortment of the parental chromosomes though crossing over. *Carothers* (1913) and others mentioned that in the metaphase I of meiosis it is usually a matter of chance whether the centromere of the paternal or the bivalent lies towards the particular pole. This results in random origin of new set

Table 13.1. Showing differences between mitosis and meiosis.

Mitosis	*Meiosis*
1. Prolonged resting stage.	1. This stage is very short.
2. The prophase is short without any substages.	2. The prophase is subdivided into six stages, viz., Proleptone, Leptonema, Zygonema, Pachynema, Diplonem and Diakinesis.
3. The chromosomes split longitudinally into sister chromatids during the preceding resting stage. From early prophase they appear as double tetrads.	3. The chromosomes do not split longitudinally.
4. The split homologous are not attracted towards each other.	4. At zygonema or succeeding stages, in prophase of first division, the homologous chromosomes are attracted towards each other and they form bivalents.
5. Synapsis and tetrad stages do not occur.	5. The chromosomes split after synapsis to from sister chromatids which are arranged in tetrads.
6. No chromosomal crossing over takes place.	6. The crossing over is prominent feature which results in exchange of genetic material.
7. The chromatids are coiled in plectonemic manner.	7. Coiling is paranemic type, and much evident.
8. Doubling of chromosomes takes place early in prophase.	8. Doubling of chromosomes occurs in pachytene stage of prophase.
9. Nucleolus and nuclear membrane disappear later on and the formation of nuclear spindle is completed with the commencement of metaphase.	9. The nuclear membrane and nucleolus disappear and the formation of acromatic figure is completed with diakinesis simultaneously.
10. The nucleus enlarge to some extent.	10. The nucleus is greatly enlarged.
11. At metaphase, the centrosome of each chromosome divides and the sister chromatids move towards the opposite poles. The somatic number of chromosomes is maintained without any reduction.	11. At metaphase I, the centrosome does not divide but the homologous chromosomes move towards opposite poles. The chromosome number is reduced to half by reduction division.
12. The chromosomes are seen in dyad consisting of two chromatids.	12. The chromosomes in the form of tetrads having four chromatids.
13. Each chromosome of paternal and maternal origin is longitudinally divided and the separation to opposite side results in identical cells at the end of mitosis.	13. At metaphase I, the orientation of chromosomes at the equatorial plate is in such pattern that either paternal or maternal chromosome goes to either pole. The orientation of one pair is independent of the other.

14. During orientation of dyads the centromeres face towards the equator and the arms directed towards the poles.	14. During the orientation of tetrads, the centromeres are placed equidistant and face the pole while arms are directed towards the equator.
15. The centromere of each bivalent undergoes division at metaphase.	15. No division of centromere occurs, but homologous chromosomes simply separate to form two dyads.
16. The initial movement of chromosomes at late metaphase is due to repulsion between the centromeric parts.	16. The main repulsive force lies between the centromeres of homologous chromosomes.
17. At anaphase the chromosomes are less thick and short.	17. The chromosomes are very thick and short.
18. At anaphase further separation of chromosomes is due to the formation of stem body.	18. Most of the changes are similar to mitosis.
19. At telophase two daughter nuclei, each with diploid number of chromosomes, are reconstituted.	19. At this phase two daughter nuclei are haploid.
20. There is no evidence of second division in mitosis.	20. There are two division for each meiotic division. The second division follows after a short gap, interkinesis. The second division is homotypic by nature, and results into four cells, in the form of tetrads.
21. In the mitosis neither the chromosomal number nor the quality is changed and the daughter cells carry 2n chromosomes.	21. At the end of meiosis each of tetrads is haploid and quality is changed.
22. This division takes place in growing somatic cells.	22. This division takes place only in gonodial cells and result in the formation of gametes.

of seperation. There may be many new combinations. In fact, since each chromosome can go to either pole, the total possible number of arrangements wlll be 2n where n is the haploid number of chromosomes

Darlington (1913) has suggested that the meiosis can be regarded as precocious mitotic, and that the force which retains daughter chromatids together in mitotic prophase also retains the homologous chromosomes pairs in the earlier stages (*Ambrose*).

Role of Centriole During Meiosis

The centrioles preform precisely controlled synchronous function during meiosis. Where meiosis occurs in two steps, that is, two nuclear divisions take place in sequence, the ctnriole reproduces and functions to from two achromatic figures. In the first division of meiosis, the centromeres do not duplicate but rather homologous chromosomes pair together and move to opposite poles, while in the second meiotic division the centromeres of chromatids are separated. A scheme for relating various key events is presented briefly as under:

1. Non-duplication of centromers in meiosis-I.
2. Non-duplication of chromosomes in meiosis-II, and
3. Duplication and function of centrioles in both divisions.

By a complete repression of the duplication, simultaneous function of the chromatic appearatus brings about the reduction. The centriolar function is super-imposed upon the duplication process of the chromatic apparatus in a sequnce which leads to the loss of a generation of chromatic apparatus. When meiosis is gametic and fertilization follows after, the immediate formation of zygote as a cell might contain the maternal as well as paternal centrioles, having endowed as it were, with the extra centriole generation. According to *Daniel Mazia* the kinetochore is an essential parts of the chromosomes so far as the mitosis is concerned. In gametogenesis there is a gain in total chromatic material, since these types of chromosomes. The extra set of centriole is rejected normally in fertilization or autogamy. Thus a cell results in a chromatic apparatus. This is accomplished by meiosis: one division meiosis gives the centriole a chance to 'catch up' in a single duplication.

14

Linkage and Genetic Mapping in Eukaryotes

According to Mendel, we expect that two different genes will segregate and independently assort themselves during gamete formation. After Mendel's work was re-discovered at the turn of the 20th century, chromosomes were identified as the cellular structures that carry genes. The chromosome theory of inheritance explained how the transmission of chromosomes is responsible for the passage of genes from parent to offspring.

When geneticists first realized that chromosomes contain the genetic material, they began to suspect that a conflict may sometimes occur between Mendel's law of independent assortment of genes and the behaviour of chromosomes during meiosis. In particular, geneticists assumed that each species of organism must contain thousands of different genes, yet cytological studies revealed that most species have at most a few dozen chromosomes. Therefore, it seemed likely (and turned out to be true) that each chromosome would carry many hundreds or even thousands of different genes. In 1911, Thomas Hunt Morgan conducted experiments showing that the transmission of genes located on the same chromosome violates the law of independent assortment.

In this chapter, we will consider the pattern of inheritance that occurs when different genes are situated on the same chromosome. We will also explore how the data from genetic crosses used to construct genetic maps that describe the order of genes along a chromosome. This chapter ends with a discussion of the eukaryotic microorganisms collectively known as fungi. This group includes yeast, such as *Saccharomyces cerevisiae* (baker's yeast), and molds, such as *Neurospora crassa* (red bread mold) and *Aspergillus nidulans* (green bread mold). We will consider the unique features of sexual reproduction that occur in certain fungal species. These features have provided the basis for distinctive genetic mapping approaches.

LINKAGE AND CROSSING OVER

In eukaryotic species, each linear chromosome contains a very long piece of DNA. As we have mentioned, a chromosome contains many individual functional units—called genes—that influence an organism's traits. A typical chromosome is expected to contain many hundreds or perhaps a few thousand different genes. The term *linkage* refers to the phenomenon that two or more genes can be located on the same chromosome. The genes are physically linked to each other, because each eukaryotic chromosome contains a single, continuous, linear piece of DNA.

As an example of linkage, let's consider a human karyotype that contains 46 chromosomes .All of the genes on each chromosome are linked to each other. Chromosomes thus are sometimes called *linkage groups*, because a chromosome contains a group of genes that are linked together. In a particular Species, there are as many linkage groups as there are types of chromosomes. For example, humans have 46 chromosomes, which are composed of 22 types of autosomes that come in pairs and one pair of sex chromosomes. There are two types of sex chromosomes, the X and Y. Therefore, humans have 22 autosomal link age groups, an X-chromosome linkage group, and a Y-chromosome linkage group.

Geneticists are often interested in the transmission of two or more traits in a genetic cross. 'When a geneticist follows two traits in a cross, this is called a *dihybrid cross*; when three traits are followed, it is a *trihybrid cross*; and so forth. The outcomes of dihybrid and trihybrid crosses depend on whether or not the genes are linked to each other along the same chromosome. In this section, we will examine how linkage affects the transmission patterns of two or more traits.

Crossing Over may Occur During the Tetrad Stage of Meiosis; When it Does, It Produces Recombinant Chromosomes and Ultimately Recombinant Phenotypes

Even though the alleles for different genes may be linked along the same chromo some, the linkage can be altered during the process of gamete formation. In diploid eukaryotic species, homologous chromosomes can exchange pieces with each other by crossing over. This event occurs frequently during prophase I of meiosis. The replicated chromosomes, known as sister chromatids, associate with the homologous sister chromatids to form a structure known as a *bivalent* or a *tetrad*. A tetrad is composed of two pairs of sister chromatids. In prophase I, a sister chromatid of one pair commonly will cross over with a sister chromatid from the homologous pair.

One of the parental chromosomes carries the *A* and *B* alleles, while the homologue carries the *a* and *b* alleles. Therefore, the gametes contain the same combination of alleles as the original chromosomes. In this case, two gametes carry the *A* and *B* alleles, and the other two gametes carry the recessive *a* and *b* alleles. The arrangement of linked alleles has not been altered.

In contrast, what can happen when crossing over occurs. Two of the gametes contain combinations of alleles (namely, *A* and *b*, *a* and *B*) that differ from those in the original chromosomes. In these two gametes, the grouping of linked alleles has been changed. An event such as this, leading to a new combination of alleles, is known as *genetic recombination*. The gametes carrying the *A* and *b*, or the *a* and *B*, alleles are called *non-parental* or *recombinant* gametes. Like wise, if these gametes participate in fertilization, the resulting offspring are called non-parental or recombinant offspring. These offspring can display combinations of traits that are different from those of either parent. In contrast, offspring that have inherited the same combination of alleles as their parents are known as *parental* or *non-recombinant* offspring.

Bateson and Punnett Discovered Two Traits that did not Assort Independently

The earliest study indicating that some traits may not assort independently was carried out by William Bateson and Reginald Punnett in 1905. As mentioned already, they were interested in the pattern of inheritance of genes in several organisms, including the chicken and the sweet pea. According to Mendel's law of in dependent assortment, a dihybrid cross between two individuals, heterozygous for two genes, should yield a 9:3:3:1 phenotypic ratio among the offspring. However, a surprising result occurred when Bateson and Punnett conducted a cross in the sweet pea involving two different traits, flower colour and pollen shape.

As seen here, they began by crossing a true-breeding strain with purple flowers (*PP*) and long pollen (*LL*) to a strain with red flowers (*pp*) and round pollen (*ll*). This yielded an F generation of plants that all had purple flowers and long pollen (*PpLl*). The unexpected result came from the F_2 generation. Even though there were four different phenotypic categories among the F_2 generation, the observed numbers of offspring did not conform to a 9:3:3:1 ratio. Bateson and Punnett found that the F_2 generation had a much greater proportion of the two parental types: purple flowers with long pollen, and red flowers with round pollen. Therefore, they suggested that the transmission of these two traits from the parental generation to the F_2 generation was somehow coupled and not easily assorted in an independent manner. However, Bateson and Punnett did not realize that this coupling was due to the linkage of the flower colour gene and the pollen shape gene on the same chromosome.

Morgan Provided Evidence for the Linkage of Several X-linked Genes and Proposed that Crossing Over between X-chromosomes can Occur

The first direct evidence that different genes can be physically located on the same chromosome came from the studies of Thomas Hunt Morgan. The earlier studies provided the groundwork to demonstrate that the genes governing X linked traits are physically linked on the X-chromosome.

Morgan investigated the inheritance pattern of many different traits that had been shown to follow an X-linked pattern of inheritance. His parental crosses were normal male fruit flies to females that had yellow bodies (*yy*), white eyes (*ww*), and miniature wings (*mm*). The wild-type alleles for these three genes are designated y^+ (gray body), w^+ (red eyes), and m^+ (normal wings). As expected, the phenotypes of the F_1 generation were wild-type females and males with yellow bodies, white eyes, and miniature wings. The linkage of these genes was revealed when the F_1 flies were mated to each other and the F_2, generation examined.

Instead of equal proportions of the eight possible phenotypes, Morgan ob served a much higher proportion of the parental combinations of traits. There were 758 flies with gray bodies, red eyes, and normal wings, and 700 flies with yellow bodies, white eyes, and miniature wings. The former combination (gray body, red eyes, and normal wings) was found in the males of the parental generation, the latter combination (yellow body, white eyes, and miniature wings) in the females of the parental generation. Morgan's proposed explanation for this higher proportion of parental combinations was that all three genes are located on the X-chromosome and, therefore, tend to be transmitted together as a unit.

However, to fully account for the data, Morgan needed to interpret two other key observations. First, he needed to explain why a significant proportion of the F_2 generation had non-parental combinations of alleles. Along with the two parental phenotypes, there were six other phenotypic combinations that were not found in the parental generation. Second, he needed to explain why there was a quantitative difference between non-parental combinations involving body colour and eye colour versus eye colour and wing length. This quantitative difference is revealed by reorganizing the data from Morgan's cross by pairs of genes:

	Total	
Gray body, red eyes	1159	
Yellow body, white eyes	1017	
Gray body, white eyes	17	Nonparental offspring
Yellow body, red eyes	12	
	2205	

	Total	
Red eyes, normal wings	770	
White-eyes, miniature wings	716	
Red eyes, miniature wings	401	Non-parental offspring
White-eyes, normal wings	318	
	2205	

We see that there were substantial differences between the numbers of non-parental offspring when pairs of genes were considered separately. It was fairly common for non-parental combinations to occur when just eye colour and wing shape were examined (401 + 318 non-parental offspring). In sharp contrast, it was rare to obtain non-parental combinations when looking at body colour and eye colour (17 + 12 non-parental offspring).

To explain these data, Morgan considered the previous studies of the French cytologist, F.A. Janssens. In 1909, Janssens proposed that crossing over involves a with his data. Overall, he made three important hypotheses to explain his results:

1. The genes for body colour, eye colour, and wing length are all located on the same chromosome (the X-chromosome), Therefore, it is most likely for all three traits to be inherited together.

2. Due to crossing over, the homologous X-chromosomes (in the female) can exchange pieces of chromosomes and create new (non-parental) combinations of alleles.
3. The likelihood of crossing over depends on the distance between two genes. If two genes are far apart from each other, it is more likely that crossing over will occur between them.

With these ideas in mind, the possible events that occurred in the F_1 female flies of Morgan's experiment. One of the X-chromosomes contained all three dominant alleles, the other all three recessive alleles. During oogenesis in the F_1 female flies, crossing over may or may not have occurred in this region of the X-chromosome. If no crossing over occurred, the parental phenotypes were produced in the F_2 offspring. Alternatively, a crossover sometimes occurred between the eye-colour gene and the wing-length gene to create non-parental offspring (namely, gray body, red eyes, and miniature wings; or yellow body, white eyes, and normal wings). According to Morgan's proposal, this is a fairly likely event, because these two genes are far apart from each other on the X-chromosome. Because of the long distance, it was fairly likely for a crossover to occur in this region. In contrast, he proposed that the body colour and eye colour genes are very close together, which makes crossing over between them an unlikely event. Nevertheless, it occasionally occurred, yielding off spring with gray bodies, white eyes, and miniature wings, or with yellow bodies, red eyes, and normal wings. Finally, it was also possible for two homologous chromosomes to cross over twice. This double crossing over is expected to be a very unlikely event. Among the 2205 offspring Morgan examined, he only found 1 fly (gray body, white eyes, normal wings) that could be explained by this phenomenon.

Chi Square Analysis can be Used to Distinguish between Linkage and Independent Assortment

Now that we have an appreciation for linkage and the production of recombinant offspring, let's consider how an experimenter can objectively decide whether two genes are linked or assort independently. Chi square analysis was introduced to evaluate the goodness of fit between a genetic hypothesis and ob served experimental data, This method is frequently used to determine if the outcome of a dihybrid cross is consistent with linkage or independent assortment.

To conduct a chi square analysis, we must first propose a hypothesis. In a dihybrid cross, the standard hypothesis is that the two genes are not linked. This hypothesis is chosen even if the observed

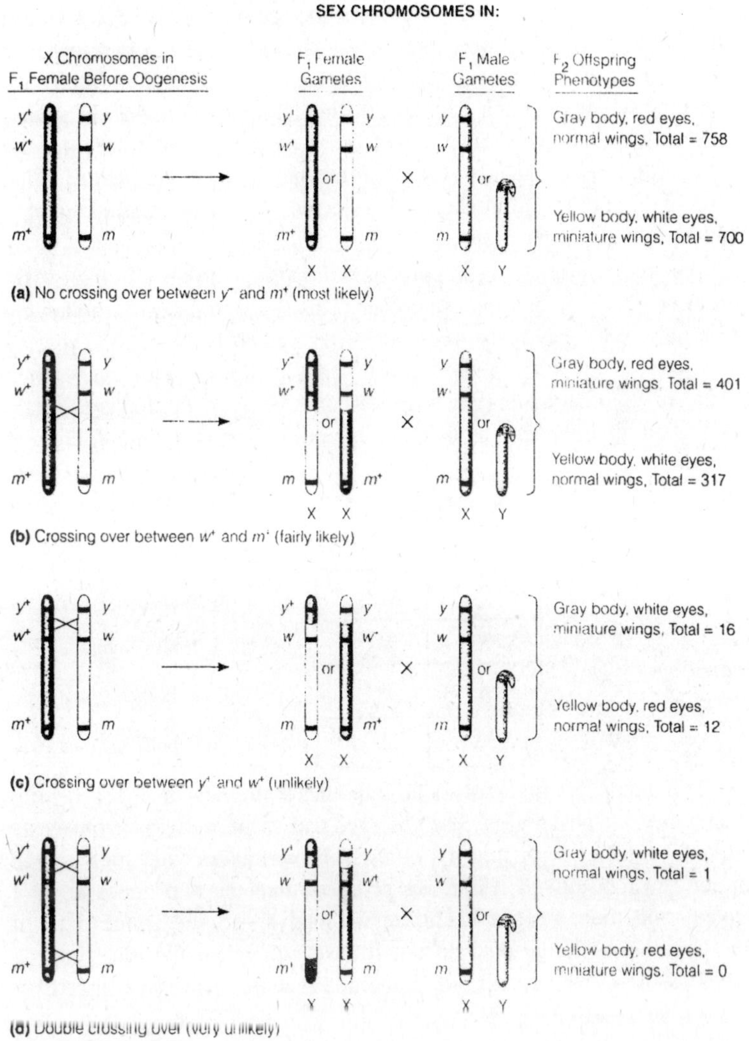

Fig. 14.1. The likelihood of crossing over provides an explanation of Morgan's trihybrid cross.

data suggest linkage, because an independent assortment hypothesis allows us to calculate the expected number of offspring based on the genotypes of the parents and the law of independent assortment. In contrast, for two linked genes that have not been previously mapped, we cannot calculate the expected number of offspring from a genetic cross, because we do not know how likely it is for a crossover to occur between the two genes. Without expected numbers of recombinant and parental offspring, we cannot use a chi square test. Therefore, we begin with the hypothesis that the genes are not linked; then, we determine whether or not our data fit this hypothesis. If the chi square value is low and we cannot reject our hypothesis, we infer that the genes assort independently. On the other hand, if the chi square value is so high that our hypothesis is rejected, we will conclude that a linkage hypothesis is correct.

As an example, let's consider the data shown on page 104 concerning body colour and eye colour. This cross produced the following offspring: 1159 gray body, red eyes; 1017 yellow body, white eyes; 17 gray body, white eyes; and 12 yellow body, red eyes. However, when a heterozygous female (y^+y w^+w) is crossed to a hemizygous male (ywY), an independent assortment hypothesis predicts the following outcome:

F$_1$ male gametes

F$_1$ female gametes	X^{yw}	Y	
$X^{y^+w^+}$	$X^{yw}X^{y^+w^+}$	$X^{y^+w^+}Y$	= Gray body, red eyes
X^{y^+w}	$X^{yw}X^{y^+w}$	$X^{y^+w}Y$	= Gray body, white eyes
X^{yw^+}	$X^{yw}X^{yw^+}$	$X^{yw^+}Y$	= Yellow body, red eyes
X^{yw}	$X^{yw}X^{yw}$	$X^{yw}Y$	= Yellow body, white eyes

The independent assortment hypothesis predicts a 1:1:1:1 ratio among the four phenotypes. The observed data mentioned above obviously seem to conflict with this hypothesis. Nevertheless, we stick to the strategy just discussed. First, we pro pose that the two genes are not linked, and then we use a chi square analysis to see if the data fit this hypothesis. If the data do not fit, we will reject the idea that the genes assort independently and conclude that the genes are linked.

An example of a chi square approach to determine linkage is shown here:

Step 1. *Propose a hypothesis.* Even though the observed data appear inconsistent with this hypothesis, we propose that the two genes for eye colour and body colour are X-linked but somehow are able to obey Mendel's law of independent assortment. This hypothesis will allow us to calculate expected values. We actually anticipate that the chi square analysis will allow us to reject the independent assortment hypothesis in favour of a linkage hypothesis.

Step 2. *Based on the hypothesis, calculate the expected values of each of the four phenotypes.* Each phenotype has an equal probability of occurring. Therefore, the probability of each genotype is 1/4. The observed F_2 generation contained a total of 2205 individuals. Our next step is to calculate the expected numbers of offspring with each phenotype when the total equals 2205; 1/4 of the offspring should be each of the four phenotypes:

$1/4 \times 2205 = 551$ (expected number of each phenotype)

Step 3. *Apply the chi square formula, using the data for the observed values (0) and the expected values (E) that have been calculated in step 2.* In this case, there are four categories within the population:

$$\chi^2 = \frac{(O_1 - E_1)^2}{E_1} + \frac{(O_2 - E_2)^2}{E_2} + \frac{(O_3 + E_3)^2}{E_3} + \frac{(O_4 - E_4)^2}{E_4}$$

$$= \frac{(1159 - 551)^2}{551} + \frac{(17 - 551)^2}{551} + \frac{(12 - 551)^2}{551} + \frac{(1017 - 551)^2}{551}$$

$$= 670.9 + 517.5 + 527.3 + 394.1$$

$$= 2109.8$$

Step 4. *Interpret the calculated chi square value.* This is done with a chi square. Since there are four experimental categories (n = 4), the degrees of freedom is $n - 1 = 3$.

The calculated chi square value is enormous! Thus, the deviation between served and expected values is very large. Such a large deviation is expected to occur by chance alone less than 1% of time. Therefore, we reject the hypothesis that the two genes assort independently. In other words, we conclude that the genes are linked.

Creighton and McClintock Correlated Crossing Over that Produced new Combinations of Alleles with the Exchange of Homologous Chromosomes

As we have seen, Morgan's studies were consistent with the hypothesis that crossing over occurs between homologous chromosomes to produce new combination of alleles. In the experiment described here, which was published two years after the work of Morgan, Harriet Creighton and Barbara McClintock an interesting strategy involving parallel observations. First, they made crosses involving two linked genes to produce parental and recombinant offspring. Second, they used a microscope to view the structures of the chromosomes in the par and in the offspring. Because the parental chromosomes had some unusual structural features, they could microscopically distinguish the two homologous chromosomes within a pair. As we will see, this enabled them to correlate the occurrence of recombinant offspring with microscopically observable exchanges in segment of homologous chromosomes.

While working in the botany department at Cornell University, Creighton and McClintock focused much of their attention on the pattern of inheritance traits in corn. In previous cytological examinations of corn chromosomes, some strains were found to have an unusual chromosome (#9) with a darkly stair knob at one end. McClintock also identified an abnormal version of this chromosome that had an extra piece of a different chromosome (#8) at the other (called an *interchange* or a *translocation*). As, this unusual version of chromosome 9 had changes at both ends that could be distinguish under the microscope.

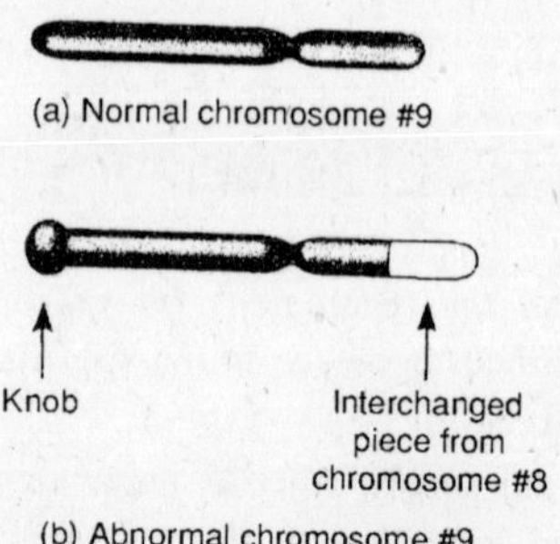

Fig. 14.2. Normal and abnormal chromosome 9 in corn used by Creghton and McClintock. A normal chromosome 9 (a) is compared with an abnormal chromosome 9 (b) that contains a knob at one end and a translocation at the opposite end.

Creighton and McClintock insightfully realized that this abnormal chromosome could be used to demonstrate that two homologous chromosomes physically exchange segments as a result of crossing over. They knew that a gene located near the knobbed end of chromosome 9 that provided colour to corn kernels. It existed in two alleles, the dominant allele *C* (coloured) and the recessive allele c (colourless). Toward the other end of the chromosome was located a second gene that affected the texture of the kernel endosperm. The dominant allele *Wx* caused starchy endosperm, while the recessive *wx* allele caused waxy endosperm Creighton and McClintock reasoned that a crossover involving a normal chromosome 9 and a knobbed/ translocated chromosome 9 would produce a chromosome that had either a knob or a translocation but not both. This chromosome would be distinctly different from either of the parental chromosomes

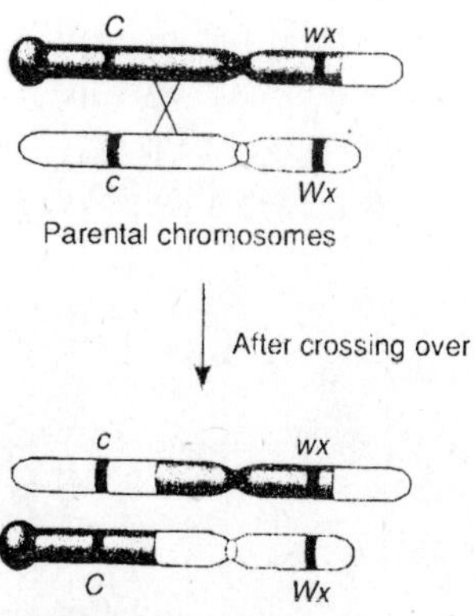

Fig. 14.3. Crossing over between normal and abnormal chromosome 9s in corn. A crossover produces a chromosome that only contains a knob at one end, and another chromosome that only contains a translocation at the other end.

Hypothesis

Offspring with non-parental phenotypes are the product of a crossover. This crossover should create non-parental chromosomes via an exchange of chromosomal segments between homologous chromosomes.

Testing the hypothesis

Starting materials: Two different strains of corn. One strain (referred to as parent A) has an abnormal #9 (knobbed/translocation) with a dominant *C* allele and a recessive *wx* allele. It also contains a cytologically normal copy of chromosome #9 that carries the recessive c allele, and the dominant *Wx* allele. Its genotype is *Cc Wxwx*. The other strain (referred to as parent B) has two normal versions of chromosome #9. The genotype of this strain is *cc Wxwx*.

1. Cross the two strains described. The tassel is the pollen-bearing structure and the silk (equivalent to the stigma and style) is connected to the ovary. After fertilization, the ovary will develop into an ear of corn.
2. Observe the kernels from this cross.
3. Microscopically examine chromosome #9 in the kernels.

Interpreting the data

In interpreting these results, we must note that Creighton and McClintock did not carry out this experiment like a standard cross, because neither of the parents were homozygous recessive for both genes. This adds some ambiguity in the relationship between the phenotypic categories and genetic recombination. In this experiment, we are interested in whether or not crossing over has occurred in parent A, which is heterozygous for both genes. This parent can produce four types of gametes, while parent B can only produce two types of gametes:

Parent A	*Parent B*
C wx (non-recombinant)	*c Wx*
c Wx (non-recombinant)	*c wx*
C Wx (recombinant)	
c wx (recombinant)	

The following types of offspring can be produced:

	c Wx	*c wx*	
C Wx	*CcWxwx* coloured/starchy	*Ccwxwx* coloured/way	Nonrecombinant
C Wx	*CcWxwx* coloured/starchy	*ccWxwx* coloured/way	Nonrecombinant
C Wx	*CcWxWx* coloured/starchy	*CcWxwx* coloured/starchy	Recombinant
c wx	*ccWxwx* coloured/starchy	*ccwxwx* coloured/way	Recombinant

Two of the phenotypic categories are ambiguous: coloured/starchy (*Cc Wxwx*) and colourless/starchy (*cc WxWx*). These phenotypes can be produced whether or not recombination occurs in parent A. Therefore, let's begin by considering the two unambiguous phenotypic categories: coloured/waxy (*Cc wxwx*) and colour less/waxy (cc wxwx). The coloured/waxy phenotype can occur only if recombination did not occur in parent A and if parent A passed the knobbed/ translocated chromosome to its offspring. As shown in the data table, three kernels were obtained with this phenotype, and all of them contained the knobbed/translocated chromosome. By comparison, the colourless/waxy phenotype can only be obtained if genetic recombination did occur in parent A and this parent passed a chromosome 9 that had a translocation but was knobless. Two kernels were obtained with this phenotype, and both of them contained the expected chromosome that had a translocation but was knobless.

Taken together, these results show a perfect correlation between genetic re combination of alleles and the cytological presence of a chromosome displaying a genetic exchange of chromosomal pieces in parent A. The other two phenotypic categories are ambiguous, because either a colourless/starchy or a coloured/starchy phenotype can be produced in the presence or the absence of genetic recombination in parent A. Nevertheless, the results agree with the hypothesis that genetic re combination is correlated with an exchange of chromosome pieces.

Overall, the observations described in this experiment are consistent with the idea that a crossover occurred, in the region between the *C* and *wx* genes, involving an exchange of segments between two homologous chromosomes. As stated by the authors, "Pairing chromosomes, heteromorphic in two regions, have been shown to exchange parts at the same time they exchange genes assigned to these regions." These results support the view that genetic recombination involves a physical exchange between homologous chromosomes. This microscopic evidence helped to convince geneticists that recombinant offspring arise from the physical exchange of segments of homologous chromosomes.

Crossing Over Occasionally Occurs During Mitosis

In multicellular organisms, the union of egg and sperm is followed by many cellular divisions, which occur in conjunction with mitotic divisions of the cell nuclei. Mitosis normally does not invoke the homologous pairing of chromosomes to form a tetrad. Therefore, crossing over during mitosis is expected to be much less likely than during meiosis. Nevertheless, it does occur on rare occasions. When it happens, mitotic crossing over may produce a pair of recombinant chromosomes that have a new combination of alleles. This is known as *mitotic recombination*. If it occurs during an early stage of embryonic development, the daughter cells containing the recombinant chromosomes will continue to divide many times to produce a patch of tissue in the adult. This may result in a portion of tissue with characteristics different from those of the rest of the organism.

In 1936, while working at the University of Rochester, Curt Stern proposed that unusual patches on the bodies of certain *Drosophila* strains were due to mitotic recombination, he was working with strains carrying X-linked alleles affecting body colour and bristle morphology. The recessive *y* allele confers yellow body colour, and the recessive *sn* allele confers shorter body bristles that look singed. The corresponding wild-type alleles confer gray body colour (y^+) and normal bristles (sn^+). Females that are y^+y sn^+sn are expected to have gray body colour and normal bristles. This was generally the case. However, when Stern carefully observed the bodies of these female flies under a low-power microscope, he occasionally noticed places in which two adjacent regions were different from the rest of the body. This is called a *twin spot*. He concluded that twin spotting was too frequent to be explained by the random positioning of two independent single spots that happened to occur close together. Instead, Stern proposed that twin spots were due to a single mitotic recombination within one cell during embryonic development.

As shown here, the X-chromosomes of the female fly are y^+ *sn* and *y* sn^+ Rarely, though, a crossover can occur during mitosis to produce two adjacent daughter cells that are y^+y^+ *snsn* and *yy* sn^+ sn^+. As embryonic development proceeds, the cell on the left will continue to divide to produce many cells, eventually producing a patch on the body that has gray colour with singed bristles. The daughter cell next to it will produce a patch of yellow body colour with normal bristles. These two adjacent patches (a twin spot) will be surrounded by cells that are y^+y sn^+sn and, thus, have gray colour and normal bristles. These infrequent twin spots provide evidence that mitotic recombination occasionally occurs.

Genetic Mapping in Diploid Eukaryotes

The purpose of *genetic mapping* (also known as gene mapping or chromosome mapping) is to determine the linear order of genes that are linked to each other along the same chromosome. Simplified genetic map of *Drosophila melanogaster* depicting the locations (i.e., loci) of many different genes along the individual chromosomes. As shown here, each gene has its own unique locus at a particular site within a chromosome. For example, the gene designated vg, which affects wing length, is located on chromosome 2. The gene designated *b*, which affects body colour, is found a moderate distance away on the same chromosome.

Even though it is an enormous amount of work, the construction of a genetic map is useful in many ways. First, it allows geneticists to understand the overall complexity and genetic organization of a particular species. The genetic map of a species portrays the underlying basis for the inherited traits that an organism displays. In some cases, the known locus of a gene within a genetic map can help molecular geneticists to clone that gene and thereby obtain greater information about its molecular features. In addition, genetic maps are useful from a evolutionary point of view. A comparison of the genetic maps among different species can improve our understanding of the evolutionary relationships among these species.

Along with these scientific uses, genetic maps have many practical benefits. For example, many human genes that play a role in human

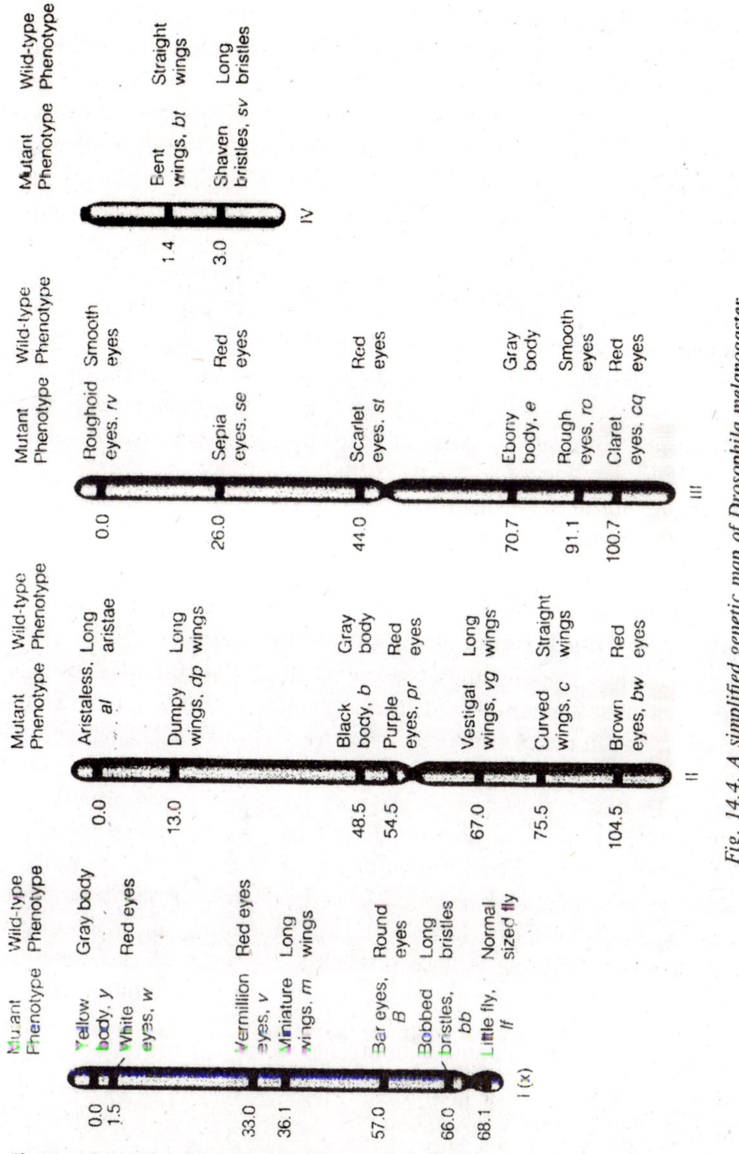

Fig. 14.4. A simplified genetic map of Drosophila melanogaster.

disease have been genetically mapped. This information can be used to diagnose and perhaps someday to treat inherited human diseases. It can also help genetic counselors predict the like hood that a couple will produce children with certain inherited diseases. In addition, genetic maps are gaining increasing importance in agriculture. A genetic map can provide plant and animal breeders with helpful information for improving agriculturally important strains through selective breeding programs.

In this section, we will begin by discussing conventional mapping techniques. These methods analyze crosses involving individuals heterozygous for two or more genes. The frequency of non-parental offspring provides a way to deduce the linear order of genes along a chromosome. This linear arrangement of genes is shown in a chart known as a *genetic linkage map*. This approach is particularly useful for analyzing organisms that are easily crossed and produce a large number of offspring in a short period of time. It has successfully mapped the genes of several plant species (with an annual generation time) and certain species of animals, such as *Drosophila*. For many organisms, however, conventional mapping approaches are difficult due to long generation times or the inability to carry out crosses (e.g., humans). Fortunately, many alternative methods of gene mapping have been developed in the past few decades. Cytological and molecular approaches also can be used to map genes.

Frequency of Recombination between Two Genes can be Correlated with their Map Distance along a Chromosome

Genetic mapping allows us to estimate the relative distances between linked genes, based on the likelihood that a crossover will occur between them. If two genes are very close together on the same chromosome, a crossover is unlikely to begin in the region between them. However, if two genes are very far apart, a crossover is more likely to be initiated in this region and thereby recombine the alleles of the two genes. Experimentally, the basis for genetic mapping is that the percentage of recombinant offspring is correlated with the distance between two genes. If two genes are far apart, many recombinant offspring will be produced. However, if two genes are close together, very few recombinant offspring will be observed.

To interpret a genetic mapping experiment, the experimenter must know if the characteristics of an offspring are due to crossing over during gamete formation in a parent. This is accomplished by conducting a *test cross*. Most test crosses are between an individual who is heterozygous for two or more genes and an individual who is recessive and homozygous for these same genes. The goal of the test cross is to determine if recombination has occurred during gamete formation in the heterozygous parent. New combinations of alleles cannot occur in the other parent, who is homozygous for these genes.

A test cross provides an experimental strategy to distinguish between recombinant and non-recombinant offspring. This cross concerns two linked genes affecting bristle length and body colour in fruit flies. The recessive alleles are *s* (short bristles) and *e* (ebony body), and the dominant (wild-type) alleles are s^+ (normal bristles) and e^+ (gray body). One parent displays both recessive traits. Therefore, we know that this parent is homozygous for the recessive alleles of the two genes (i.e., *ss ee*). The other parent is heterozygous for the linked genes affecting bristle length and body colour. This parent was produced from a cross involving a true-breeding wild-type fly and a true-breeding fly with short bristles and an ebony body. Therefore, in this heterozygous parent, we know that the *s* and *e* alleles are linked on one chromosome and the corresponding s^+ and e^+ alleles are linked on the homologous chromosome.

Now let's take a look at the four possible types of offspring these parents can produce. The offspring's phenotypes are normal bristles/ gray body, short bristles/ebony body, short bristles/gray body, and normal bristles/ebony body. All four types of offspring have inherited a chromosome carrying the s and e alleles from their homozygous parent (shown on the right in each pair). Focus your attention on the other chromosome. The offspring with short bristles and ebony bodies have also inherited a second chromosome carrying the *s* and *e* alleles from their other parent. This chromosome is not the product of a crossover in the heterozygous parent. The offspring with normal bristles and gray bodies have inherited a chromosome carrying the s^+ and e^+ alleles from the heterozygous parent. Again, this chromosome is not the product of a crossover.

The other two types of offspring, however, can only be produced if crossing over has occurred. Those with normal bristles and ebony bodies or short bristle and gray bodies have inherited a chromosome that is the product of a crossover during gamete formation in the heterozygous parent. As noted in, the recombinant offspring are fewer in number than are the non-recombinant offspring

The data shown at the bottom of can be used to estimate the distance between the two genes. The map distance is defined as the number of recombinant offspring divided by the total number of offspring, multiplied by 100. With the data, we can calculate the map distance between the *s* and *e* alleles using this formula:

$$\text{Map distance} = \frac{\text{Number of recombinant offspring}}{\text{Total number of offspring}} \times 100$$

$$= \frac{76 + 75}{542 + 537 + 76 + 75} \times 100$$

$= 12.3$ map units

The units of distance are called *map units* (mu) or sometimes *centimorgans* (cM) in honor of Thomas Hunt Morgan. In this example, we would say that the s and e alleles are 12.3 map units apart from each other along the same chromosome.

Alfred Sturtevant Used the Frequency of Crossing Over between Two Genes to Produce the First Genetic Map in 1911

The first individual to construct a (very small) genetic map was Alfred Sturtevant, an undergraduate who spent time in the laboratory of Thomas Hunt Morgan. In 1965, more than fifty years after he constructed the first genetic map, Sturtevant wrote: In the latter part of 1911, in conversation with Morgan . . . I suddenly realized that the variations in the strength of linkage, already attributed by Morgan to differences in the spatial separation of the genes, offered the possibility of determining sequences [different genes] in the linear dimension of a chromosome. I went home and spent most of the night (to the neglect of my undergraduate homework) in producing the first chromosome map, which included the sex-linked genes, y, w, v, m, and r, in the order and approximately the relative spacing that they still appear on the standard maps.

In the experiment described here, Sturtevant considered the outcome of crosses he conducted involving six different mutant alleles that altered the phenotype of normal flies. All of these alleles were known to be recessive and X-linked. They are *y* (yellow body colour), *w* (white eye colour), *w-e* (eosin eye colour), *v* (vermilion eye colour), *m* (miniature wings), and *r* (rudimentary wings). The *w* and *w-e* alleles are alleles of the same gene. In contrast, the *v* allele (vermillion eye colour) is an allele of a different gene that also affects eye colour. The two alleles that affect wing length, *m* and *r*, are also in different genes. Therefore, Sturtevant studied the inheritance of six recessive alleles, but since *w* and *w-e* are alleles of the same gene, his genetic map only contained five genes. The corresponding wild-type alleles are y^+ (gray body), w^+ (red eyes), v^+ (red eyes), m^+ (normal wings) and r^+ (normal wings).

Hypothesis

When genes are located on the same chromosome, the distance between the genes can be estimated from the proportion of recombinant offspring. This provides a way to map the order of genes along a chromosome.

Testing the hypothesis

Starting materials: Sturtevant began with several different strains of *Drosophila* that contained the six alleles already described.

1. Cross a female that is heterozygous for two different genes to a male that is hemizygous recessive for the same two genes. In this example, cross a female that is $X^{y+w+}X^{yw}$ to a male that is $X^{yw}Y$. This strategy was employed for many dihybrid combinations of the six alleles already described.
2. Observe the outcome of the crosses.
3. Calculate the percentage of offspring that are the result of crossing over (# of non parental/total).

Data

Alleles concerned	*Number recombinant/ total number*	*Percent recombinant offspring*
y and *w/w-e*	214/21,736	1.0
y and *v*	1,464/4,551	32.2
y and *r*	115/324	35.5
y and *m*	260/693	37.5
w/w-e and *v*	471/1,584	29.7
w/w-e and *r*	2,0632/6,116	33.7
w/w-e and *m*	406/898	45.2
v and *r*	17/573	3.0
v and *m*	109/405	26.9

Interpreting the data

Let's begin by contrasting the results between particular pairs of genes. In some dihybrid crosses, the percentage of non parental offspring was rather low. For example, dihybrid crosses involving the *y* allele and the *w* or *w-e* allele yielded 1% recombinant offspring. This result suggests that these two genes are very close together. By comparison, other dihybrid crosses showed a higher percentage of non-parental offspring. Crosses involving the *v* and *m* alleles produced 26.9% recombinant offspring. These two genes are expected to be farther apart.

To construct his map, Sturtevant began with the assumption that the map distances would be more accurate between genes that are closely linked. Therefore, his map is based on the distance between *y* and *w* (1.0), *w* and *v* (29.7), *v* and *r* (3.0), and *v* and *m* (26.9). He also considered other features of the data to deduce the order of the genes. For example, the percentage of crossovers between *w* and *r* was 33.7. The percentage of crossovers between *w* and *v* was only 29.7, suggesting that *v* is between *w* and *r*, but closer to *r*. The proximity of *v* and *r* is confirmed by the low percentage of crossovers between *v* and *r* (3.0). Sturtevant collectively considered all these data and proposed the genetic map shown here:

In this genetic map, Sturtevant began at they allele and mapped the genes from left to right. For example, the *y* and *v* alleles are 30.7 map units apart, and the *v* and *r* alleles are 3.0 (mu) apart. This study by Sturtevant was a major breakthrough, since it showed how to map the locations of genes along chromosomes by making the appropriate crosses.

If you look carefully at Sturtevant's data, you will notice that there were two observations that do not agree very well with his genetic map. The percentage of recombinant offspring for the *y* and *m* dihybrid cross was 37.5 (but the map distance is 57.6), and the crossover percentage between *w* and *m* was 45.2 (but the map distance is 56.6). As the percentage of recombinant off spring approaches a value of 50%, this value becomes a progressively more inaccurate measure of map distance. When the distance between two genes is large, the likelihood of multiple crossovers in the region between them causes the observed number of recombinant offspring to underestimate this distance. In addition, multiple crossovers set a quantitative limit on the relationship between map distance and the percentage of recombinant offspring. Even though two different genes can be on the same chromosome and more than 50 map units apart, a test cross is only expected to yield a maximum of 50% recombinant offspring.

Trihybrid Crosses can be Used to Determine the Order of and Distance between Linked Genes

Until now, we have been considering the construction of genetic maps using dihybrid test crosses to compute map distance. The data from trihybrid crosses can also yield information about map distance and gene order. In a trihybrid cross, the experimenter crosses two individuals that differ in three traits. The following experiment follows a common strategy for using trihybrid crosses to map genes. In this experiment, the parental generation consists of fruit flies that differ in body colour, eye colour, and wing shape. We must begin with true-breeding lines so that we know which alleles are initially linked to each other on the same chromosome. In this ex ample, all the dominant alleles are linked on the same chromosome.

Step 1. *Cross two true-breeding strains that differ with regard to three alleles.* In this example, we will cross a fly that has a black body (*bb*), purple eyes (*prpr*), and vestigial wings (*vgvg*) to a homozygous wild-type fly ($b^+b^+pr^+pr^+vg^+vg^+$): The goal in this step is to obtain F_1 individuals that are heterozygous for all three alleles. In the F_1 heterozygotes, all the dominant alleles are located on one chromosome, all the recessive alleles on the other homologous chromosome.

Step 2. *Perform a test cross by mating F_1 female heterozygotes to male flies that are homozygous recessive for all three alleles (bb prpr vgvg)*: During gametogenesis in the heterozygous female F_1 flies, crossing over may occur to produce new combinations of the three alleles.

Step 3. *Collect data for the F_2 generation.* There are eight possible phenotypic combinations:

Phenotype	*Number of observed offspring*
Gray body, red eyes, normal wings	411
Black body, purple eyes, vestigial wings	412
Gray body, purple eyes, vestigial wings	30
Black body, red eyes, normal wings	28
Gray body, red eyes, vestigial wings	61
Black body, purple eyes, normal wings	60
Gray body, purple eyes, normal wings	2
Black body, red eyes, vestigial wings	1

Analysis of the F_2 generation flies will allow us to map these three genes. Since the three genes exist as two alleles each, there are $2^3 = 8$ possible combinations of offspring. If these alleles assorted independently, all eight combinations would occur in equal proportions. However, we see that the proportions of the eight phenotypes are far from equal.

The genotypes of the parental generation correspond to the phenotypes gray body, red eyes, and normal wings and black body, purple eyes, and vestigial wings. In crosses involving linked genes, the parental phenotypes occur most frequently in the offspring. The remaining six phenotypes are due to crossing over.

Two of the phenotypes (namely, gray body, purple eyes, and normal wings; and black body, red eyes, and vestigial wings) arise from a double crossover between two combinations of genes. The double crossover is always expected to be the least frequent category of offspring. Also, the combination of traits in the double crossover tells us which gene is in the middle. When a chromatid undergoes a double crossover, it separates the gene in the middle from the other two genes at either end. In the double crossover categories, the recessive purple-eye allele is separated from the other two recessive alleles. When mated to a homozygous recessive fly in the test cross, this yields flies with gray bodies, purple eyes, and normal wings, or with black bodies, red eyes, and vestigial wings. This observation indicates that the gene for eye colour lies between the genes for body colour and wing shape.

Step 4. *Calculate the map distance between pairs of genes.* To do this, we must regroup the data according to pairs of genes. From the parental generation, we know that the dominant alleles are initially linked to each other, as are the recessive alleles. This allows us to group pairs of genes into parental and non-parental combinations. The parental combinations are composed of a pair of dominant or a pair of recessive genes, whereas non-parental combinations have one dominant and one recessive gene. After we have regrouped the data in this way, the map distance between two genes can be calculated:

Parental offspring	*Total*	*Non-parental offspring*	*Total*
Gray body, red eyes,		Gray body, purple eyes,	
(411 + 61)	472	(30+2)	32
Black body, purple eyes,		Black body, red eyes	
(412 + 60)	472	(28+1)	29
	944		61
Gray body, normal wings		Gray bodies, vestigial	
(411 + 1)	413	(30+61)	91
Black body, vestigial wings		Black body, normal wings	
(412 + 28)	413	(28+60)	88
	826		179
Red eyes, normal wings		Red eyes, vestigial wings	
(411 + 28)	439	(61 + 1)	62
Purple eyes, vestigial wings		Purple eyes, normal wings	
(412 + 30)	442	(60 + 2)	62
	881		124

Map distance between body colour and eye colour:

$$\text{Map distance} = \frac{61}{944 + 61} \times 100 = 6.1 \text{ mu}$$

Map distance between body colour and wing shape :

$$\text{Map distance} = \frac{179}{826 + 179} \times 100 = 17.8 \text{ mu}$$

Map distance between eye colour and wing shape :

$$\text{Map distance} = \frac{124}{881 + 124} \times 100 = 12.3 \text{ mu}$$

Step 5. *Construct the map*. Based on the map unit calculation, the body colour and wing shape genes are farthest apart. The eye colour gene must lie in the middle. As mentioned earlier, this order of genes is also confirmed by the pattern of traits found in the double crossovers. To construct the map, we use the distances between the genes that are closest together.

In our example, we have placed the body colour gene first and the wing shape gene last. Our data also are consistent with a map in which the wing shape gene comes first and the body colour gene comes last. In detailed genetic maps, the locations of genes are mapped relative to the centromere.

Interference can Influence the Number of Double Crossovers that Occur in a Short Region

The product rule allows us to predict the expected likelihood of a double crossover provided we know the individual probabilities of each single crossover. Let's reconsider the data of the trihybrid test cross just described to see if the frequency of double crossovers is what we would expect based on the product rule. If we multiply the likelihood of a single crossover between *b* and *pr* (6.1%) times the likelihood of a single crossover between *pr* and *vg* (12.3%), then the product rule predicts

Expected likelihood of a double crossover = 0.061 × 0.123 = 0.0075 = 0.75%

Based on a total of 1005 offspring produced,

Expected number of offspring due to a double crossover = 1005 × 0.0075 = 7.5

In other words, we would expect about 7 or 8 offspring to be produced as a result of a double crossover. The observed number of offspring was only 3 (namely, 2 with gray bodies, purple eyes, and normal wings, and 1 with a black body, red eyes, and vestigial wings).

This lower than expected value is not due to random sampling error. Instead, it is due to a common genetic phenomenon known as *positive interference*. When a crossover occurs in one region of a chromosome, it often decreases the probability that another crossover will occur nearby. In other words, the first crossover interferes with the ability to form a second crossover in the immediate vicinity. To provide interference with a quantitative value, we first calculate the *coefficient of coincidence* (C):

$$C = \frac{\text{Observed number of double crossover}}{\text{Expected number of double corssover}}$$

Interference (I) is expressed as

$$I = 1 - C$$

For the data of the trihybrid test cross, the observed number of crossovers is 3 and the expected number is 7.5, and so the coefficient of coincidence equals 3/7.5 = 0.40. In other words, only 40% of the expected number of double crossovers were actually observed. The value for interference equals 1 - 0.4 = 0.60 or 60%. This means that 60% of the expected number of crossovers were prevented from occurring. Since *I* has a positive value, this is positive interference. Rarely, the out come of a test cross yields a negative value for interference. A negative interference value suggests that a first crossover enhances the rate of a second crossover in a nearby region. Although the molecular mechanisms that cause interference are not entirely understood, most organisms regulate the number of crossovers so that very few occur per chromosome. The reasons for positive and negative interference will require further research.

Genetic Mapping in Haploid Eukaryotes

Before ending our discussion of genetic mapping, it is interesting to consider some pioneering studies that involved the genetic mapping of haploid organisms. It may be surprising to you that certain species of lower eukaryotes, particularly unicellular algae and fungi, which spend the greatest part of their life cycle in the haploid state, have also been used in mapping studies. The sac fungi (ascomycetes) have been particularly useful to geneticists because of their unique style of sexual reproduction. In fact, much of our earliest understanding of genetic recombination came from the genetic analyses of fungi.

Fungi may be unicellular or multicellular organisms. Fungal cells are typically haploid (1*n*) and can reproduce asexually. In addition, fungi can also reproduce sexually by the fusion of two haploid cells to create a diploid zygote (2*n*). The diploid zygote can then proceed through meiosis to produce four haploid cells, which are called *spores*. This group of four spores is known as a *tetrad* (not to be confused with a tetrad of four sister chromatids). In some species, meiosis is followed by a mitotic division to produce eight cells, known as an *octad*. The cells of a tetrad or octad are contained within a sac known as an *ascus* (plural, asci). In other words, the products of a single meiotic division are contained within one sac. This is a key feature that is useful to geneticists, and it dramatically differs from sexual reproduction in animals and plants. For example, in animals, oogenesis produces a single functional egg, and spermatogenesis occurs in the testes, where the resulting sperm become mixed with millions of other sperm.

Using a microscope, researchers can dissect asci and study the traits of each haploid spore. In this way, these organisms offer a unique opportunity for geneticists to identify and study all of the cells that are derived from a single meiotic division. In this section, we will consider how the analysis of asci can be used to map genes in fungi.

In Fungal Asci, Tetrads and Octads of Spores can be Ordered or Unordered

The arrangement of spores within an ascus varies from species to species. In some cases, the ascus provides enough space for the tetrads or octads of spores to randomly mix together. This is known as an *unordered tetrad* or *octad*. These occur in fungal species such as *S. cerevisiae* and *A. nidulans* and also in certain unicellular algae (*Chlamydomonas rheinhardii*). By comparison, other species of fungi produce a very tight ascus that prevents spores from randomly moving around. This can create a *linear tetrad* or *octad*. In this example, spores which carry the A allele have orange pigmentation, while spores having the a (albino) allele are white.

A key feature of linear tetrads or octads is that the position and order of spores within the ascus reflects their relationship to each other as they were produced by meiosis and mitosis. After the original diploid cell has undergone chromosome replication, the first meiotic division produces two cells that are arranged next to each other within the sac. The second meiotic division then produces four cells that are also arranged in a straight row. Due to the tight enclosure of the sac around the cells, each pair of daughter cells is forced to lie next to each other in a linear fashion. Likewise, when each of these four cells divides by mitosis, each pair of daughter cells is located next to each other.

Linear Tetrad Analysis can be Used to Map the Distance between a Gene and the Centromere

In the case of species that make linear tetrads and octads, experimenters can analyze the phenotypes of the spores within the asci and map the distance between a single gene and the centromere. Since the location of the centromere can be seen under the microscope, the mapping of a gene relative to the centromere provides a way to correlate a gene's location with the cytological characteristics of a chromosome. This approach has been extensively exploited in *N. crassa.*

The arrangement of cells within a *Neurospora* ascus depending on whether or not a crossover has occurred between two homologues that differ at a gene with alleles *A* (orange pigmentation) and *a* (albino, which results in a white phenotype). The octad contains a linear arrangement of four haploid cells carrying the *A* allele, which are adjacent to four haploid cells that contain the *a* allele. This 4:4 arrangement of spores within the ascus is called a *first-division segregation* (FDS) *pattern* or an *M1 pattern*.

In contrast, if a crossover occurs between the centromere and the gene of interest, the linear octad will deviate from the 4:4 patter Depending on the relative locations of the two chromatids that participated in the crossover, the ascus will contain a 2:2:2:2 or 2:4:2 pattern. These are called *second-division segregation* (SDS) *patterns* or *M2 patterns*.

Since a pattern of second division segregation is a result of crossing over, the percentage of M2 asci can be used to calculate the map distance between the centromere and the gene of interest. To understand why this is possible, let's consider the pattern of movement of a crossover site, or *chiasma* (plural, *chiasmata*). A., a chiasma forms at a particular site on a chromosome and then moves away from the centromere and toward the terminal end of the chromosome. As shown here, a crossover will only separate a gene from its original centromere if it begins in the region between the centromere and that gene. Therefore the chances of getting a 2:2:2:2 or 2:4:2 pattern depend on the distance between the gene of interest and the centromere.

To determine the map distance between the centromere and a gene, the experimenter must count the number of SDS asci and the total number of asci. In SDS asci, only half of the spores are actually

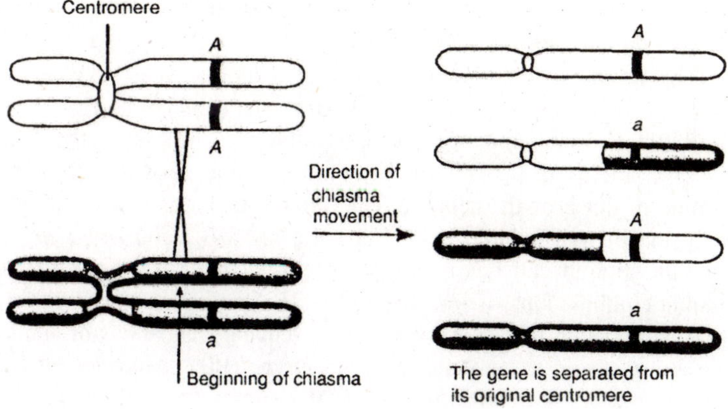

(a) Chiasma begins between centromere and gene of interest

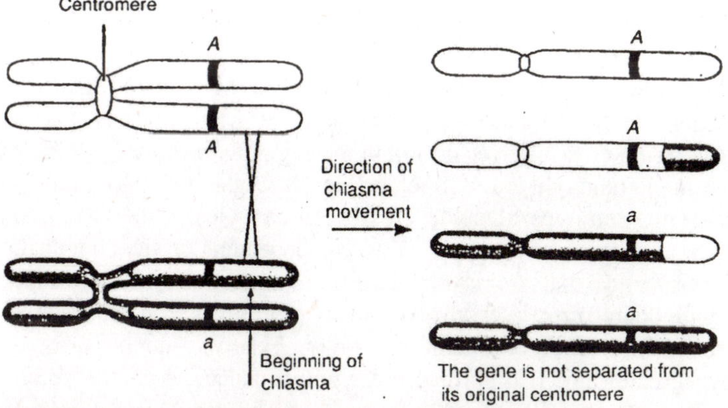

(b) Chiasma does not begin between centromere and gene of interest

Fig. 14.5. The movement of chiasmata during crossing over.

the product of a crossover. Therefore the map distance is calculated as

$$\text{Map distance} = \frac{(1/2)\ (\text{Number of SDS asci})}{\text{Total number of asci}} \times 100$$

Unordered Tetrad Analysis can be Used to Map Genes in Dihybrid Crosses

Unordered tetrads contain a group of spores that are randomly arranged and the product of meiosis. An experimenter can conduct a dihybrid cross, remove the spores from each asci, and determine the phenotypes of the spores. This analysis can determine if two genes are linked or assort independently. If two genes are linked, a tetrad analysis can also be used to compute map distance.

The possible outcomes starting with a diploid yeast zygote that has the genotype *ura*$^+$*ura*-2 *arg*$^+$*arg*-3. *Ura*$^+$ and *arg*$^+$ are normal alleles required for uracil and arginine biosynthesis. *Ura*-2 and *arg*-3 are defective alleles that result in yeast strains that require uracil and arginine in the growth medium. This diploid cell was produced from the fusion of two haploid cells that were *ura*$^+$*arg*$^+$ and *ura*-2 *arg*-3. After the diploid cell has completed meiosis, there are three distinct possible combinations of four haploid cells. One possibility is that the tetrad will contain four spores with the parental combinations of alleles. This ascus is said to have the *parental ditype* (PD). Alternatively, an ascus with a *non-parental ditype* (NPD) contains four cells with non-parental genotypes. Finally, it is possible to have a ascus that has two parental cells and two non-parental cells. This is called a *tetratype* (T).

When two genes assort independently, the number of asci having a parental ditype is expected to equal the number having a non-parental ditype, thus yielding 50% recombinant spores. For linked genes, the relationship between crossing over and the type of ascus that will result. If no crossing over occurs in the region between the two genes, then the parental ditype will be created. A single crossover event will produce a tetratype. Double crossovers can yield either a parental ditype, tetratype, or non-parental ditype depending on the combination of chromatids that are involved. A non-parental ditype is produced when a double crossover involves all four chromatids. A tetratype will result from a three chromatid crossover. Finally, a double crossover between the same two chromosomes will produce the parental ditype.

The data from a tetrad analysis can be used to calculate the map distance between two linked genes. As in conventional mapping, the map distance is calculated the percentage of offspring that carry recombinant chromosomes. As mentioned, a tetratype contains 50% recombinant chromosomes, a non-parental ditype 100%. Therefore, the map distance is computed as

$$\text{Map distance} = \frac{\text{Nonparental ditypes} + (1/2)\ (\text{Tetratypes})}{\text{Total number of asci}} \times 100$$

Over short map distances, this calculation provides a fairly reliable measure of distance. However, it does not adequately account for double crossovers. When two genes are far apart on the same chromosome, the calculated map distance using this equation underestimates the actual map distance due to multiple crossing over. Fortunately, a particular strength of tetrad analysis is that we can derive another equation that accounts for double crossovers and thereby provides a more accurate value for map distance. To begin this derivation, let's consider a more precise way to calculate map distance:

$$\text{Map distance} = \frac{\text{Single crossover tetrad} + (2)\,(\text{Double crossover tetrad})}{\text{Total number of asci}} \times 0.5 \times 100$$

This equation includes the number of single and double crossovers in the computation of map distance. The total number of crossovers equals the number of single crossovers plus two times the number of double crossovers. Overall, the tetrads that contain single and double crossovers also contain 50% non-recombinant chromosomes. To calculate map distance, therefore, we divide the total number of crossovers by the total number of asci and multiply by 50%.

Next, we need to relate this equation to the number of parental ditypes, non-parental ditypes, and tetratypes that are obtained by experimentation. To derive this relationship, we must consider the types of tetrads that are produced from no crossing over, a single crossover, and double crossovers. As shown there, the parental ditype and tetratype are ambiguous. The parental ditype can be derived from no crossovers or a double crossover; the tetratype can be derived from a single crossover or a double crossover. However, the non-parental ditype is unambiguous, since it can only be produced from a double crossover. We can use this observation as a way to determine the actual number of single and double crossovers. Therefore, the total number of double crossovers equals 4 times the number of NPD.

Next, we need to know the number of single crossovers. A single crossover will yield a tetratype, but double crossovers can also yield a tetratype. Therefore, the total number of tetratypes overestimates the true number of single crossovers. Fortunately, we can compensate for this overestimation. Since there are two types of tetratypes that are due to a double crossover, the actual number of tetratypes arising from a double crossover should equal 2NPD. Therefore, the true number of single crossovers is calculated as T – 2NPD.

Now we have accurate measures of both single and double crossovers. The number of single crossovers equals T – 2NPD, and the number of double crossovers equals 4NPD. We can substitute these values into our previous equation:

$$\text{Map distance} = \frac{(\text{T} - 2\ \text{NPD}) + (2)\,(4\ \text{NPD})}{\text{Total number of asci}} \times 0.5 \times 100$$

$$= \frac{\text{T - 6 NPD}}{\text{Total number of asci}} \times 0.5 \times 100$$

This equation provides a more accurate measure of map distance, since it considers both single and double crossovers.

Linkage refers to the phenomenon that many different genes may be located the same chromosome. Chromosomes are sometimes called *linkage groups* because they contain a group of linked genes. Linkage affects the pattern of inheritance, because closely linked genes do not assort independently during gamete formation. This produces a greater percentage of offspring that display parental phenotypes. Nevertheless, non-parental offspring can be produced as result of *crossing over*.

The likelihood of crossing over depends on the *distance* between two genes. If two genes are far apart from each other on the same chromosome, it is more likely that crossing over will occur between them. Therefore, when two genes are widely separated, a substantial percentage of recombinant offspring will be obtained fro a *test cross*. However, the percentage of recombinant offspring cannot exceed value of 50%, even when two genes are more that 50 map units apart on the same chromosome. The relationship between the percentage of recombinant offspring and the linear distance between genes is the basis for *genetic mapping*.

Experimentally, the phenomenon of linkage was deduced from genetic crosses. Bateson and Punnett were the first scientists to notice that certain genes do not as sort independently. Morgan conducted crosses involving X-linked traits in fruit flue and correctly proposed that linkage is due to the location of particular genes on the same chromosome. He also hypothesized that recombinant phenotypes occur be cause of crossing over during meiosis. Morgan realized that the likelihood of crossing over depends on the distance between two genes. The proposal that genetic recombination is due to crossing over was confirmed cytologically by the studies of Creighton and McClintock, which showed that the production of recombination offspring correlates with the production of recombinant chromosomes.

Genetic mapping is the determination of gene order and distance along chromosomes. In this chapter, we have considered how test crosses are conducted as method to map genes. Sturtevant was the first person to understand that the percentage of recombinant offspring in a test cross could be used as a measure of the relative distance between two genes. Map distance is computed as the number of recombinant offspring divided by the total number of offspring. This approach can be readily applied to map genes using dihybrid and trihybrid test crosses. Genetic mapping is most accurate when map distances are calculated between closely linked genes. As the map distance approaches 50 map units (mu) and above, the percent age of recombinant offspring is not a reliable measure of map distance.

This chapter ended with a discussion of gene mapping methods in fungi. I group of fungi known as the ascomycetes have been extensively used in genetic studies, because they produce all the products of a single meiosis within an ascus For fungi such as *Neurospora* that make a linear ascus, the spores are arranged in manner that reflects their relationship to each other during meiosis (and mitosis) Linear asci can be analyzed to map the location of a single gene relative to the centromere. Fungal species that produce unordered asci have also been used in map ping studies. In this case, dihybrid crosses are made, and the distance between the two genes can be computed by determining the proportions of parental ditypes tetratypes, and non-parental ditypes. Although bacteria normally reproduce asexually, they still can transfer genetic material by various different mechanisms. As we will see, these mechanisms also provide a way to map genes along the bacterial chromosome.

15

CYTOLOGICAL GENETICS

The field of cytological genetics has grown up at the interface of molecular genetics and cytology, or cell biology. The structures of cells as they relate to cell division and gene expression are studied by *cytogeneticists*. With regard to the first main gene function, self-duplication or reproduction, the structure and function of chromosomes and the spindle are particularly relevant. With regard to the second main gene function, phenotypic expression, the structure and function of interphase chromatin and of the nucleolus are important.

CHROMOSOME BANDING

Several techniques for staining mitotic or meiotic chromosomes, so that characteristic transverse bands can be seen, were developed in the early 1970s. The methods and results of some of the major techniques are given in Three of the techniques shown (*G banding*, *C banding*, and *R banding*) are usually accomplished by staining pretreated chromosomes with the same dye mixture, Giemsa stain. This dye results in coloured (dark) bands and light interbands. The fourth technique, *Q banding*, is performed using quinacrine, a fluorescent dye. The stained bands are then fluorescent when observed using UV microscopy.

The mechanisms underlying these techniques have been investigated, and, although there is not complete agreement, many cytogeneticists now think that both Giemsa and quinacrine stain with DNA rather than the chromosomal proteins. G banding seems to result from stacking of dye molecules along the sides of certain regions of the DNA. The interbands may stain less because more of their DNA is covered by proteins and/or because more of the interbands DNA is extracted before stain is applied. The G-bands have been identified as regions that are

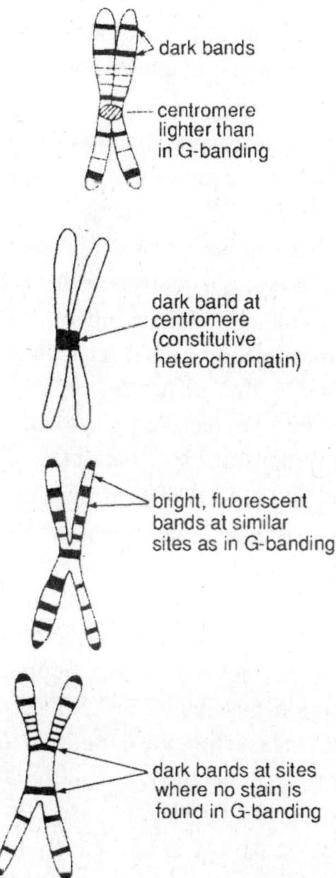

Fig. 15.1. Effects of major mitotic chromosome banding techniques on a hypothetical chromosome.

heterochromatic; these same regions replicate late in the S period of the cell cycle and thus can be differentially labelled using 3 added at the end of S.

Another interesting correlation is between G bands of mitotic chromatin and the dark staining clumps of meiotic chromatin called *chromomeres*. During the early prophase of meiosis 1 (pachytene), the chromosomes are condensed but not maximally contracted. The arrangement of chromomeres in such chromosomes is the same as the arrangement of C bands in mitotic chromosomes of the same organism. In meiotic prophase, the extended human chromosomes can be seen to have even more chromomeres than G bands (2000 to 300 chromomeres rather than 350 G bands listed in the Paris nomenclature), but as the chromosomes condense further, the 350 characteristic bands are produced by coalescence of several chromomeres to form each band. Certain manipulations of mitotic metaphase chromosomes have revealed a similar underlying complexity.

The C-banding technique involves stringent pretreatments, which extract the DNA from the chromosome except for regions of constitutive heterochromatin. The Giemsa stain then binds to this remaining DNA. Some data indicate that depurination (release of purine bases, A and G, leaving the sugar-phosphate back bone intact) may be essential to C banding. The regions stained by C banding are usually stained less intensely than in the G-banding technique (which uses the same stain, as mentioned). This difference in the effects of the same stain may result from a more effective protein cover over centromeric DNA during G banding than during C banding.

The Q banding technique is based on the fact that AT-rich DNA enhances the fluorescence of quinacrine, whereas GC-rich DNA quenches the fluorescence. Q bands correspond in most cases to G bands, thus suggesting heterochromatin is the major site of Q banding. Like G banding, Q banding produces less centromeric staining than expected. Probably, protein-DNA interactions at this region limit dye access or limit the type of interaction possible. The Q banding technique is very useful in detecting and measuring the length of the Y chromosome, since the Y fluoresces very brightly with this stain. A number of observations have suggested that the Y varies in length in different men, and that this length variability is exclusively in the heterochromatin of the Y. The easy identification of the Y with Q banding has greatly facilitated such studies.

R banding or reverse banding is thought to result in Giemsa staining of GC rich DNA, especially noticeable in phase-contrast microscopy. Other stains (for example, acridine orange) work equally well in producing R bands after the same pretreatment. One possible mechanism for R banding involves denaturation and extraction of the AT-rich DNA, leaving behind the GC-rich DNA, which is more resistant to thermal denaturation. R bands are, in general, those regions that do no stain well in G banding or Q banding. They are thus euchromatin. These bands have been shown to replicate early in S periods.

The characteristics of DNA from the R bands (euchromatin, expressed genes), the G bands (*facultative heterochromatin*, inactive genes), and the C bands (*constitutive heterochromatin*, highly repetitive non Genic DNA). The model shows how the chromosome might be organized into G bands and inter bands (which would stain during R

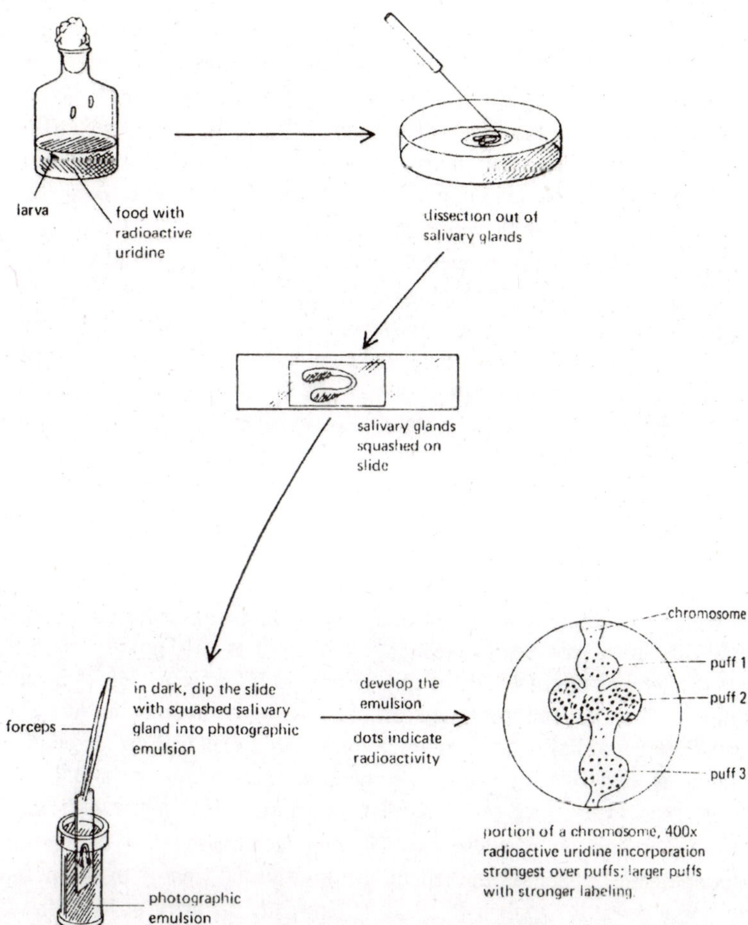

Fig. 15.2. The process of demonstrating enhanced transcription of RNA in puffs via autoradiography after radioactive uridine incorporation.

banding). Notice how both helical and clumped levels of organization are used to build up the final condensed chromosome, according to this model.

Structure and Functions of Special Types of Chromatin

The role of nonhistone proteins in opening our the structure of these chromosomes for gene expression was also covered. At that point, *puffs* along the polytene chromosomes were merely identified as sites of RNA transcription. This finding was worked out by cytogeneticists in a series of experiments that will now be described. In 1959, Pelling showed that puffed regions of *Chironomus* polytene chromosomes corresponded to sites of active incorporation of radioactive uridine, an RNA precursor. Soon after that Clever and Karlson and others showed that ecdysone (an insect steroid hormone) elicits some of the puffing sequence characteristic of normal development, implying that steroid control was being exerted via gene transcription (RNA synthesis) control.

The technique of *in situ* hybridization, developed by Pardue and Gall, has extended the scope of analysis of RNA transcription from polytene chromosomes. Pardue and collaborators have recently compared the sites of labelled uridine incorporation (a transcription autoradiogram) with the sites to which extracted labeled RNA from the same tissue will hybridize (*in situ* hybridization) for the *Drosophila* salivary glands. The chromosomes have been stained to reveal structure and the dark dots indicate sites of radioactivity.

There is a strong but not a perfect correlation between sites where RNA is made and sites where RNA hybridizes. Note that 74EF and 75B are puffs that label heavily in either case, indicating RNA synthesis and homology to RNA product. Note also that non-puffed regions (bands) as well as puffs incorporate uridine. Bands that transcribe but do not hybridize have not been detected, although it is difficult to be sure whether or not minor bands of this sort could be present. The opposite finding, that is, the presence of bands that hybridize but do not transcribe, is relatively common. A reasonable explanation for such a finding is that the band contains some DNA homologous to a DNA in a different, transcribed puff, but that the band itself is not transcribed.

A comparison of these two labeling techniques as a bar graph plotted against chromosome length. An unexpected finding is that puffs do not account for most hybridization sites; the majority of hybridized sites are in unpuffed bands. It is significant that high levels of *in situ* hybridization in puffs are found upon hybridization with RNA probes labeled during transcription of those same puffs. For bands that puff, puffing thus does seem to go hand-in-hand with transcription.

A different sort of refinement of our understanding of puff structure comes from the observation that in vivo hybridization does not always label the entire puff uniformly, as ^{3}H-uridine incorporation does. This sort of finding in the 75B region, where the ^{3}H-uridine covers the puff but the labeled RNA hybridizes at only one end of the puff. This finding may indicate that much of the RNA product of the puff is lost during RNA processing an idea consistent with regulation via RNA processing.

The use of fluorescent antibodies to localize proteins in the polytene chromosomes. This technique has also revealed that RNA polymerase is heavily concentrated in puffs, although it is also present in some bands. Another special type of chromosome is the lamp brush chromosome. This chromosomal structure is seen during the late pachytene and diplotene of meiotic I prophase in oocytes. RNA synthesis occurs along the loops of these chromosomes, while the DNA along the axis is inactive. RNA synthesis is characteristically observed during meiotic prophase I, even when the lamp brush configuration cannot be detected. A typical pattern of events is that autoradiography after ^{3}H-uridine incorporation in zygotene and pachytene stages shows the whole chromosome complement to be involved in RNA synthesis, whereas later, in the diplotene stage of meiotic I prophase, rRNA is almost the only type of RNA made, By metaphase I, RNA synthesis is undetectable. It is thus reasonable to assume that lampbrush structure is the reflection of a normal activity of meiotic prophase in many organisms, namely RNA transcription.

A paired set of lampbrush chromosomes, held together by two chiasmata, and a view of the RNA synthesis in progress along one of the loops. In *Triturus cristatus* the *lampbrush loops* persist about 200 days. Gall and Callan showed that in one giant loop, transcription begins where the end of the loop reaches the axis of the chromosome, and progresses around the loop. It takes about 10 to 14 days of continuous labeling with ^{3}H-uridine for the label to proceed all the way around the loop. When the labeled RNA is found at the thick end of the loop, it is evidently about to be released from the chromosome as ribonucleoprotein. The thick (RNA rich) and thin (RNA poor) ends of such a loop, as well as evidence for the progression of ^{3}H-uridine labelling around the loop. It is probable that such RNA is an important set of mRNAs needed during early development of the fertilized egg.

A third special type of chromosome, not previously described, is the B chromosome. *B chromosomes* are extra chromosomes that are not vital to the organism. They may be present or absent in different members of a population without conferring any obvious advantages or disadvantages. Such chromosomes exist in hundreds of species of plants and animals, yet since they have little or no effect on the phenotype. They have not been extensively studied. In mealy bugs, radioactive RNA was synthesized from organisms having only the normal (A set) of chromosomes. Then, cells with both A and B chromosomes were incubated with the labeled RNA for *in situ* hybridization. The B chromosomes were lightly labeled or unlabeled, whereas the A chromosomes were heavily labeled. This experiment suggests that B chromosomes have little DNA homology with A chromosomes.

In corn, the B chromosome is denser than the regular A chromosomes at mitotic metaphase. In addition, it replicates late in the mitotic S period, probably because it is highly heterochromatic. This B chromosome controls a system of non-Mendelian inheritance, involving non disjunction at the second pollen mitosis followed by preferential fertilization of the egg by sperm containing B chromosomes. (*Non-disjunction* is the failure of a pair of daughter chromatids to separate during cell division.) Evidently, the corn B chromosome also enhances recombination in certain regions of the A chromosomes. It has been suggested that the corn B chromosome may have structural features in common with abnormal 10, an unusual corn chromosome consisting of a very heterochromatic region attached to chromosome 10.

Detection of Changes in Chromosome Number and Structure

Change in Number Chromosomes

Changes in chromosome number may create aneuploid organisms. *Aneuploids* have changed in chromosome number to a number of chromosomes not divisible by a, the haploid number. In some organisms there is a supernumerary B set of chromosomes, as described in the previous section. In other organisms (and in the A set of chromosomes, if both A and B sets are present) there is a necessity for a balanced set of chromosomes. The loss of a chromosome, representing loss of several hundred to several thousand genes, represents a loss of gene copies which may be needed at times of maximal gene expression. Such loss is expected to have a deleterious effect. It is more surprising to discover that gain of a chromosome has deleterious effects, although they are usually less severe than in chromosome loss. The explanation must lie in the necessity to have an optimum balance between the gene products encoded by the different types of chromosomes.

One mechanism producing aneuploidy is non-disjunction. If non disjunction occurs during meiosis, a gamete with two copies of that chromosome is formed as well as a gamete with no copies. Fertilization by such gametes leads (barring developmental arrest) to entire aneuploid organisms. Non-disjunction can also happen later in development, during a mitotic cell division. This situation produces a mosaic organism, in which most cells are normal but those descended from the cell with abnormal mitosis are aneuploid. If sex chromosomes are involved in the non disjunction, the genotypes produced are X0, XXY, and so on. Deficiency of one autosome is called *monosomy*, whereas the presence of an extra autosome is called *trisomy*, by contrast with the normal diploid state with two copies of each autosome. Non-disjunction occurs in many organisms, including humans.

In 1959, Lejeune, Gautier, and Turpin showed that Down's syndrome (mongolism) due to trisomy of one of the two shortest human chromosomes (21 or 22). Down's syndrome had been recognized medically since the mid-1800s. Symptoms of this trisomy include mental deficiency, short stature, a round face with epicanthal eyelid folds, and a simian fold pattern of lines in the palm of the hand. Since identical twins in which Down's syndrome occurred were nearly always either both mongoloid or neither mongoloid, it was assumed to be hereditary. The puzzling feature of the inheritance was that only very rarely was more than one person affected in a particular family. Establishing that the condition was due to an extra autosome convincingly explained these findings. Using G banding, one can easily distinguish the two shortest pairs of autosomes. When chromosome banding became widely used, the extra chromosome of Down's patients was called number 21, in accordance with the literature traditions.

Few individuals with other trisomies survive until birth in humans. Trisomies of 13 and 18 have been described, and both produce a variety of effects including mental retardation. Nonviable individuals with trisomies for other chromosomes do occur, however, as has been shown by examining the chromosomes of spontaneously aborted features.

The sex chromosomes can also undergo non-disjunction, and many persons with abnormal doses of sex chromosomes are nearly normal in most phenotypic qualities except for sexual differentiation. The mildness of such effects is probably due to the inactivity of all but one X in adult organisms and the presence of very few genes on the Y chromosome. Two of the most common syndromes involving sex chromosome aneuploidy are Klinefelter's (XXY or other rarer forms such as XXXY) and Turner's (X0) syndromes. Patients with *Klinefelter's syndrome* are phenotypically male but with underdeveloped genitalia, occasional breast development and less body hair than usual. The incidence is approximately one (XXY) male per 1000 births. They may or may not be mentally deficient. The XXY configuration poses problems in reproduction, and most such people are sterile. Patients with Turner's syndrome are phenotypically female but with underdeveloped ovaries. The incidence is about one (X0) per 20,000 births, but 99% of X0 conceptions. Persons with *Turner's syndrome* are usually short and have underdeveloped secondary characteristics, but are often normal mentally. They may have reduced mathematical and spatial ability.

A third common abnormality is the XYY occurring approximately once per 1000 births. This syndrome is rather controversial, there having been some studies suggesting increased aggressiveness from XYY men and others refuting this notion. The XYY males are fertile and seem normal in most phenotypic characteristics. They have some tendency to be extra tall and possibly somewhat reduced in intelligence. Other syndromes, much rarer, have been described in humans with the following sex chromosome sets: XXX, XYYY, XXXX, XXXXX, XXXXY, XXYY, and XXXYY.

Detection of Chromosomal Rearrangements by Different Chromosome Shapes, Sizes, or Banding Patterns

In addition to changes in the number of chromosomes, a number of macrolesions have been described that can be detected (a) by different banding patterns from normal, suggesting that chromosomal rearrangements have occurred, (b) by meiotic synapsis of particular chromosomes showing that homologous regions have been rearranged, or (c) by polytene chromosome synapsis showing different arrangements of the same bands.

A great deal can be told about normalcy of a chromosome complement by simply examining a *karyotype* (all the metaphase chromosomes from a cell). A human genetic disease called the cri-du-chat syndrome is characterized by severe mental deficiency, a very round face, microcephaly and a very plaintive cry in infancy. In 1963, Lejeune and coworkers showed that a partial deletion of the short arm of chromosome 5 was responsible for this syndrome. This deletion essentially converts the submetacentric chromosome to an acrocentric, so it is easily visible in the karyotype.

The *Bar* eye mutation in *Drosophila*, which was shown, by examining the salivary gland chromosomes, to result from a duplication of a short set of bands and interbands. This mutation is only one of many duplications and deficiencies which have been detected by examining *Drosophila* chromosome bands.

In addition, the mitotic banding techniques for human chromosomes have permitted detection of rearrangements. Q banding of chromosomes of a person with Down's syndrome but with only the normal number

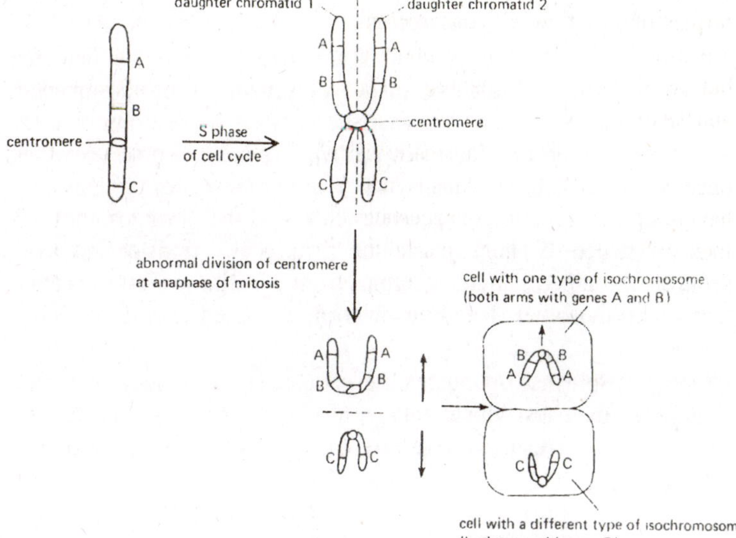

Fig. 15.3. Formation of isochromosomes by division of the centromere in the wrong plant.

of chromosomes, 46. Careful examination of this figure will show that a chromosome 21 has been translocated to the short arm of one of the chromosomes 14. In addition, 15-21 translocations are often implicated in cases of Down's syndrome. Another human abnormality distinguishable by banding is a variant of Turner's syndrome (Xiso X), caused by chromosomal aberration in a person with 46 chromosomes. One of the X chromosomes in such a patient is an *isochromosome* (both arms carry the same genes) for the long arm of the X chromosome. Consequently, no genes commonly found on the short arm of X are present on this chromosome; thus, the patient has only one copy of the X short arm genes. Isochromosomes are thought to form by division of the centromere in the wrong plane.

In view of the currently accepted unineme model of the chromosome, breakage and reunion of the DNAs involved would presumably be necessary. The Philadelphia chromosome associated with chronic myelocytic leukemia is another translocation that can be detected by banding. The usual type of aberrant chromosome observed is a translocation from chromosome 22 to chromosome 9. The persons affected are essentially always mosaic for the very short 22, the Ph´ chromosome, suggesting that the translocation occurred in a somatic cell rather late in development.

Detection of Rearrangements by Meiotic Synapsis or Interphase Synapsis of Polytene Chromosomes

Many types of chromosomal rearrangements can be detected because they cause distinctive shapes of synapsed chromosomes in prophase of meiosis I. The characteristic synapsis pattern observed for a *reciprocal translocation* (in which two types of chromosomes exchange pieces) is a cross-shaped structure with four chromosomes participating, in terms of the genetic consequences. Two of the chromosomes are normal in structure and two are the reciprocally translocated pair. Such a cross-shaped pattern resulting from a reciprocal translocation between chromosomes 8 and 10 of corn, synapsed during meiosis I prophase.

One possible use of such observations is in identifying which chromosome matches which linkage group (defined by transmission genetics). Such correlations have been made in Neurospora, the orange bread mold, by Barry. A summary of part of his study is a stock having a reciprocal translocation between linkage groups I and II showed a group of four chromosomes synapsed in pachytene, and the cytologically identifiable chromosomes involved were 1 and 6, ranked in order of size. These and other similar data using other translocation heterozygotes established linkage group I to be chromosome 1, linkage

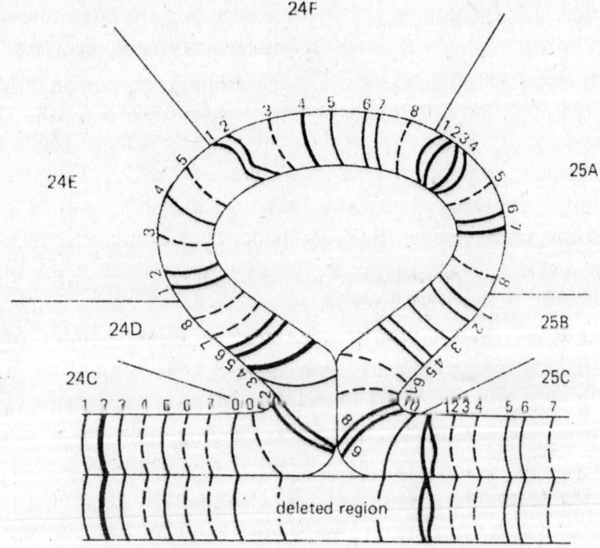

Fig. 15.4. Diagram of a loop formed in somatically paired polytene chromosome in a region of a deletion in chromosome II of Drosophila.

group II to be chromosome 6, and linkage group VII to be chromosome 7. In addition to details of banding and meiotic synapsis, a third way of cytologically defining chromosomal rearrangements is via the somatic synapsis found for the polytene chromosomes. In a heterozygote for a chromosomal rearrangement, such pairing is quite characteristic in shape, depending on the type of aberration. For example, shows a looped synapsis in an inversion heterozygote. What happens in a deletion heterozygote. (Note that the loop starts and ends at the same place along the deleted chromosome, but has been stretched for facility in diagramming. Synapsis is thus a powerful tool for detecting rearrangements, whether it is observed in meiotic prophase or in polytene somatic cells.)

Ultrastructure of Mitotic and Meiotic Chromatin

Mitotic Chromatin

Electron microscopy of dividing cells has given us a mental image of a single chromosome at the start of metaphase. In chromosomes sectioned and observed by transmission electron microscopy (TEM) at metaphase, not much structure is usually apparent in the DNA-protein complex, but the points of attachment of the spindle fibers can be seen. The scanning electron microscopy (SEM) process has generated a rather different view of mitotic chromosomes, also represented in the drawing of Wolfe. This view shows a knobby structure (with *microconvules* or knobs consisting of supercoiled loops) typified.

An interesting question with regard to mitotic chromatin is how it is condensed from the extended nucleosome state to the 300Å

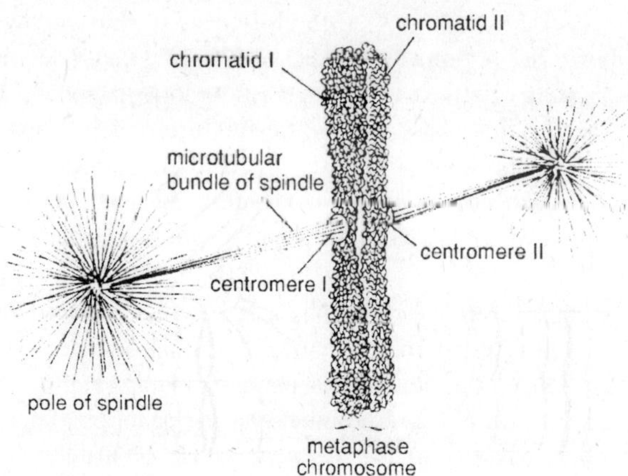

Fig. 15.5. Mitosis, showing a chromosome at mitotic metaphase.

nucleohistone fiber and finally to the knobby chromosome. Laemmli and coworkers have approached this question by removing the histones from metaphase chromatin on the hypothesis that higher order structure is stabilized by means of nonhistone proteins, and one might find out how DNA is associated with structural nonhistone proteins if those two components were observed in the absence of histones. A chromosome of this type, isolated from a HeLa cell. The dark material in the center is a nonhistone protein *scaffold*, and the DNA is spilling out on all sides. The DNA has no visible free ends, suggesting that DNA loops might have both ends rooted at the scaffold. The view in such loops in a region of less dense DNA, convincingly demonstrating that the scaffold has associated with it both ends of these loops, at least in the region shown.

In the absence of histone extraction, the looped DNA would have been in the nucleosome or nu-body structure. This structure is packed in some way to form the 300-Å diameter unit fiber of chromatin. Until recently, the suggestion had been, almost universally, that the nucleosomes formed a regular spiral of 300 Å diameter. Presently, there is some evidence that this unit fiber may be an association of clumps of nucleosomes, and that these clumps may vary in size. The 200- to 300 Å (20- to 30-nm) fibers from a chicken erythrocyte nucleus, showing apparent variation along the fiber axis in the degree of nucleosome packing.

Meiotic Chromatin

In order to discuss meiotic chromatin, it is necessary to describe the stages of meiotic divisions I and II especially the substages of prophase I. As meiosis begins, each chromosome shortens by coiling. Its centromere, the region for spindle attachment, may become visible as a constriction. This earliest stage of meiotic prophase is called *leptotene*. During this phase, a specific mixture of RNA and proteins called *synaptinemal complex* (SC) forms between the two sister chromatids of each chromosome. Just before the next phase, this SC moves from between the chromatids to cover one surface of each chromatid.

You might visualize each chromosome at this point as an opened out hot-dog bun, spread over the whole inside surface with peanut butter. Each half of the bun represents a *sister chromatid*, the whole bun represents the whole chromosome, and the peanut butter represents the SC in this analogy. At the *zygotene* phase of prophase, the two homologs of each chromosome stick together by means of their SC

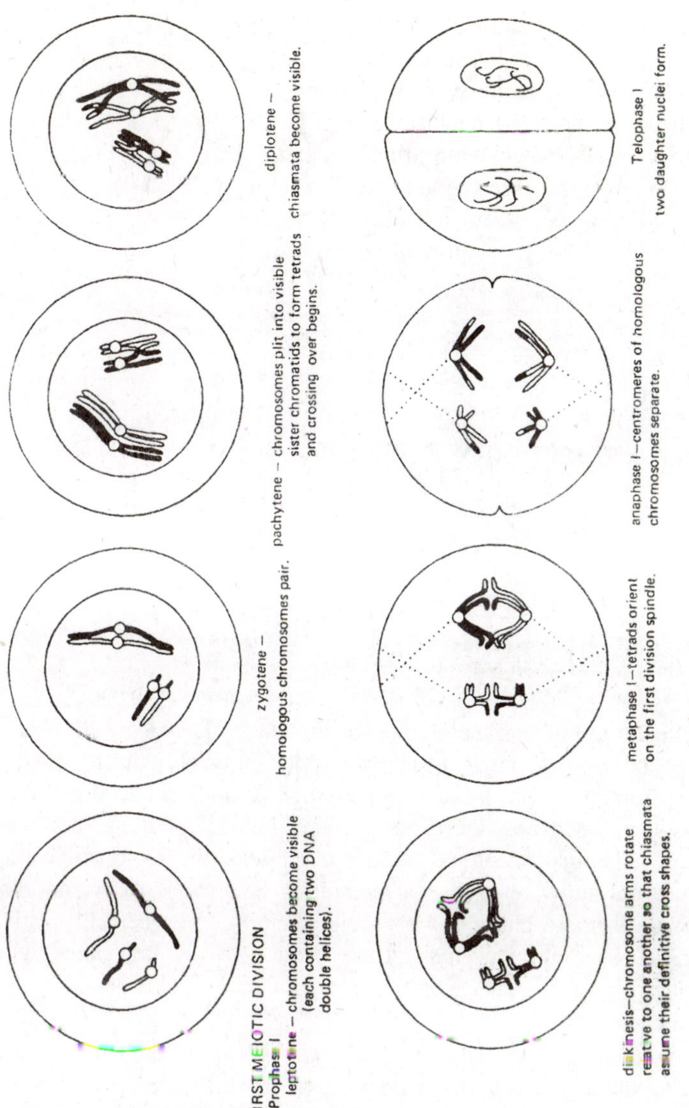

Fig. 15.6. Stages of meiosis, showing major cytological events.

surfaces. In this analogy, we would take another exactly similar peanut-butter-spread, opened-up hot-dog bun and stick the two buns together by their peanut butter surfaces. This situation of lengthwise pairing of homologous maternal and paternal chromosomes is called *synapsis*. Notice that both chromatids of each homolog are paired with their matching chromosome all along their lengths.

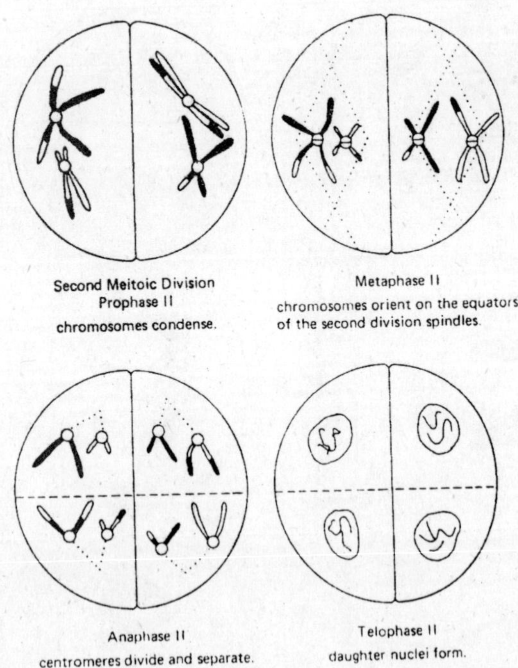

Fig. 15.7. Stages of meiosis, showing major cytological events.

During the next stage of meiotic prophase, the *pachytene* phase, each of the two homologs in each pair splits into visibly distinguishable sister chromatids. The set of four paired chromatids is called a *tetrad*. Meiosis will separate these four strands into separate gametes or progeny. The analysis of such tetrads is a fascinating subject revealing much about how genes are arranged. It is during pachytene that chiasmata form. A chiasma is a visible connection between chromatids of homologous chromosomes. These *chiasmata* are correlated with crossing over, the breakage and exchange of homologous pieces by the paired chromatids.

The following two stages, *diplotene* and *diakinesis*, are characterized by an apparent repulsion of the homologs away from each other, except at the centromeres and the chiasmata. As the chromosomes move apart, the chiasmata *terminalize* (move toward the ends of the chromosomes). Then prophase of meiosis I is complete. Some cells (for example, cells defined as human egg cells) remain frozen at this stage for years. At metaphase I, the chromosomes align themselves along the equator of the spindle so that the maternal homolog's centromere faces one pole of the spindle and the paternal homolog's centromere faces the other pole. Notice that there is no sorting mechanism to make all of the maternal chromosomes face one pole and the paternal chromosomes face the other.

Now, in *anaphase I*, the homologs separate. In telophase, the two cells separate. Commonly, each of these cells receives some chromosomes from each parent. We have shown each cell resulting from meiosis I receiving a paternal homolog of one chromosome and a maternal homolog of the other type of chromosome, but which homolog goes where is merely a matter of chance. In an organism with only two types of chromosome, often one might see complete separation of maternal from paternal chromosomes at this point. In humans, with 23 pairs of chromosomes, it is almost inconceivable ($p = (0.5)^{23} = 1/8,388,608$) that a cell produced by meiosis would get a complete set of chromosomes from only one of the two parental sets.

The next two phases, *telophase I* and *prophase II*, may be abbreviated or absent. They are mainly concerned with decondensation and recondensation of the chromosomes. In *metaphase II*, the chromosomes orient on the metaphase plate so that the centromeres of the sister chromatids of each of the chromosomes face opposite poles. The centromeres then separate in *anaphase II* and each chromatid ends up in a separate cell from its sister chromatid. In *telophase II*, the chromosomes decondense and meiosis is complete. Four cells result, but all four may not be viable entities. In egg development only one of the four is typically viable, and this one receives almost all of the cytoplasm.

Meiotic prophase I (recall from earlier in this section that the stages are leptotene, zygotene, pachytene, diplotene, and diakinesis) has been examined extensively by cytologists in order to find out the nature of synapsis and the nature of crossing over. As already described, the synaptinemal complex (SC) appears to be a necessary part of synapsis. Evidence for this conjecture comes from studies of *Drosophila*, where the males do not exhibit meiotic crossing over but the females do have crossing over. The females in this case form SC and have chiasmata along their chromosomes during meiosis I prophase, but the males have no SC and are achiasmatic. A mutant, $c(3)G^{17}$ has no crossing over in the females and also has no SC. We can infer that chromosomes without SC probably cannot undergo crossing over.

Although the SC is important in chiasma formation, it is not sufficient to insure chiasmata. In the silk moth (*Bombyx mori*), both males and females have SC, but females are achiasmatic. The structure of the SC in females becomes modified during pachytene, and these modifications may prevent recombination. Another example of a developed SC without the presence of chiasmata is the Black Beauty hybrid lily, in which Hotla and Stern have shown that an important endonuclease, necessary to break the homologous DNA strands for recombination, is missing.

A synapsed pair of chromosomes, held together by the SC, is usually anchored to the nuclear envelope on both ends. In some organisms, these attached ends are widely scattered over the nuclear envelope, whereas in others they form a bouquet at a single, small area on the nuclear envelope. An electron micrograph of the end of a pachytene synapsed pair attached to the nuclear envelope, and a study tracing exactly where all the chromosome ends were anchored in a *Locusta* pachytene spermatocyte (note that all ended on one side of the nuclear envelope). Although this type of structure is found in most organisms examined, there are exceptions. In *Drosophila* for example, the chromosomes are not attached to the nuclear envelope.

When pairing and synapsis begin, there are few coils holding the pair together along its length; however, the process of synapsis during zygotene is accompanied by a coiling, possibly produced by rotation of one or both of the anchored ends. By mid-pachytene, these coils are observed throughout the length of the pair, unless aberrations (for example, translocations) interfere with their formation. In a rat spermatocyte, by late pachytene the total number of such counter-clockwise 1800 twists, counting all the paired chromosomes, can approach 100 turns. In very late pachytene, the number of these coils usually decreases. This figure also shows that the SC is forming between the homologs at a region removed from the centromere.

Zickler has studied electron microscopic images of synapsed chromosomes in order to determine what the appearance of a chiasma is at a higher level of resolution. In the fungus, *Sordaria macrospora*, there are seven chromosomes. Each long chromosome pair has three to four chiasmata, each middle-sized pair has two to three chiasmata, and the short pair has one chiasma. The average number of chiasmata per nucleus, in 20 nuclei scored using light microscopy, was 18. The structure on the electron microscopic level, which corresponded in quantity, was the *synaptinemal complex node* (SC node). There were 17 of these in one pachytene nucleus and 21 in another. There were from one to four nodes present in any synapsed pair of chromosomes. Such nodes have been described in a number of other organisms as well, particularly by Moens. An electron micrograph of synapsed rat chromosomes in which nodes appear as definite, discrete, very darkly staining regions, often extending across the space occupied by the SC itself. In other organisms, the size, shape, and location of such structures may differ. In all cases studied, though, a modification of the SC structure appears a likely candidate for recombination sites.

The special lampbrush chromosomes of meiotic prophase in many oocytes have been characterized by cytologists in terms of the centromere placements, the placement of certain large loops, the placement of knobs of chromatin, the placement of chromatin-associated proteinaceous granules, and the location of nucleoli. An example of such a map, as worked out for *Triturus marmoratus*. It is generally possible to recognize such chromosomes by use of relative lengths and these working maps of conspicuous features.

The main feature of meiosis subsequent to prophase I that is of interest is the centromere placement insuring proper alignment at metaphase I and metaphase II a drawing of the alignment at metaphase I, and electron micrographs showing the contrast in centromere alignments in metaphases I and II.

DNA Replication and Transcription at the Cytological Level

Data implying that bacterial DNA replicates semiconservatively, as obtained by autoradiography of entire *E. coli* chromosomes. In addition, the eyelike openings where DNA replication occurs in eukaryotic chromosomes have been. The semiconservative nature of whole chromosome replication remains to be described. In 1957, Taylor and coworkers labeled chromosomes with 3 during one cell generation, then followed the distribution of label in subsequent generations. Some results from their study, demonstrating that the label was passed on exactly as predicted, assuming unineme chromosomes with semiconservative replication of DNA. Notice that some new and old segments have been interchanged. The extent of sister-strand exchange was not known at that time, but it is rather extensive.

Transcription of RNA has been studied extensively in polytene and in lampbrush chromosomes. In addition to transcription from special chromosomes, transcription from RNA- and mRNA-coding regions of DNA has been observed via electron microscopy.

Foe, for example, studied the electron microscopic appearance of chromatin in milkweed bug embryos after various lengths of

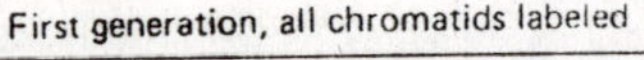

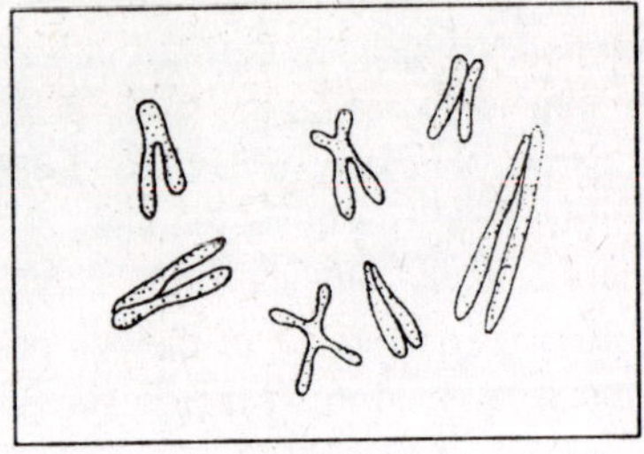

Second generation

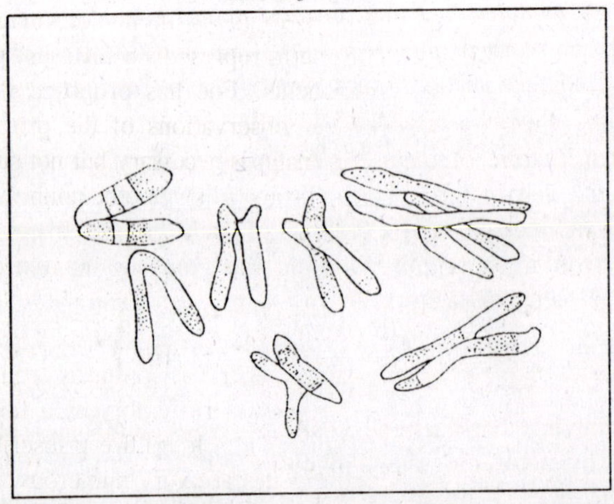

Fig. 15.8. Chromosome semiconservative DNA replication in Bellevalia romana.

development. At 68 hr of development, the neurula stage, rRNA is being synthesized very actively. A typical tandemly repeated array of rRNA genes at this stage contained 51 fibers (transcription product) with transcribed regions of the DNA averaging 2.4 μm long, separated by untranscribed spacers averaging 0.6 μm long. The entire group of fibers being transcribed from one 2.4 μm long region of DNA was termed an rTU, *ribosomal transcription unit*. By 116 hr of development, rRNA transcription has been partially deactivated. There are only about 13 fibers per rTU, and, in fact active rTUs occur less frequently Since this lower fiber number results in a less densely packed "brush" on the rTU, it is possible to see whether or not the chromatin is beaded into nucleosomes under the transcription products. The answer seems to be that it is not beaded.

The rDNA has an average width of 73 Å, where DNA is 20 Å in diameter and a nucleosome is 100 Å. Since micrococcal nuclease-

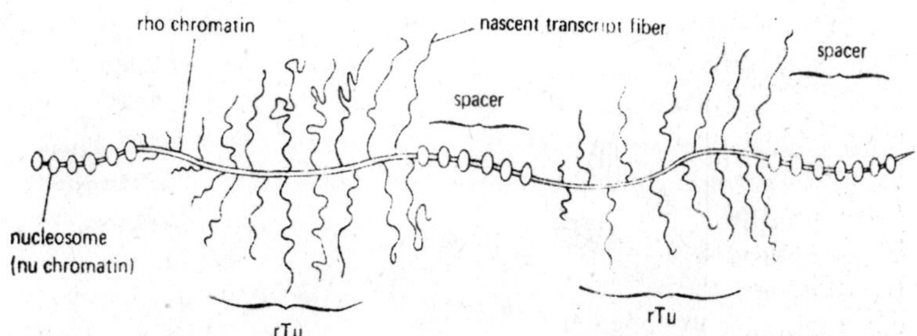

Fig. 15.9. Diagram of ribosomal transcription units (rTU), showing nascent transcript fibers and the spacers between rTU.

sensitive sites are spaced similarly to those in nucleosomes, it is probable that the non-beaded chromatin represents a different packing of the general nucleosome components. Foe has proposed that this state be termed *rho* chromatin. Her observations of the progress of development suggested that *rho* chromatin is necessary but not sufficient for rDNA transcription, since *rho* chromatin's smooth, nonbeaded 70-A fibers are present in some cases before any nascent fibers can be detected. Both the number of active rTUs and the level of transcription of single units appear to be modulated during development.

McKnight and associates, on the other hand, studied the ***Drosophila** melanogaster* embryo nonribosomal gene transcription as well as the ribosomal gene transcription. In preblastoderm embryos, a few short regions of nonribosomal DNA (0.5 to 3 μm long) are transcribed. In general, RNA polymerases on these genes are very numerous, leading to dense brushes of nascent RNA or ribonucleoprotein fibrils. The rDNA is not active in preblastoderm embryos of the fruit fly.

The rDNA is transcribed as the embryo reaches the cellular blastoderm stage (when a syncitial structure is partitioned into cells). These rTUs are not activated simultaneously, and may be regulated independently by different signal molecules. Non-rDNA cistrons, similar to those present before cellular blastoderm formation, continue to be transcribed. In addition, there is transcription of regions with very low RNA polymerase frequency and without clear cut gradients in lengths of nascent fibrils. McKnight and coworkers found both sparsely and densely RNA polymerase populated non-ribosomal regions had beaded structure like the nucleosomes. Antibodies to the histones reacted to similar extents with the beads of transcribed and nontranscribed chromatin, implying that both types of structure contain histones. But techniques would not permit saturation level antibody reactions, so the meaning of this observation is not certain.

McKnight and coworkers examined the relationship between replication and transcription. The replication evidently does not disrupt nucleosome structure significantly. A number of electron micrographs of the cellular blastoderm embryos, replicating and transcribing actively, have shown transcriptional gradients on newly replicated chromatin. Both daughter chromatids in such a gradient have nascent ribonucleoprotein particles. With regard to rRNA genes, when both replication forks of the replication eye can be seen, it appears that the origin of replication was located in a non-transcribed spacer. The replication complex in some cases dislodges RNA polymerase and nascent rRNA or ribonucleoprotein, but in other cases it may follow the RNA polymerase. Patterns that can be interpreted in either of these fashions have been seen.

Sister Chromatid Exchange

The two chromatids that are produced during chromosome replication are called sister chromatids. Until recently, it was not possible to distinguish the sister chromatids from each other, so the level of exchanges between sister chromatids in somatic cells could not be assessed cytologically. A debate raged for years over the extent of sister strand exchange in synapsed pairs of chromosomes during meiosis. Conclusive proof that ring chromosomes underwent *sister chromatid exchanges* (SCE) in both meiosis and mitosis was presented by McClintock and by Schwartz before the 1960s. Evidently, ring chromosomes were felt to be sufficiently atypical that results regarding them might not apply to all chromosomes.

In the late 1950s and early 1960s, Taylor studied chromosome replication, and found that it was possible to detect SCEs via ^{3}H-thymidine labeling techniques. When he prevented the separation of cells by adding coichicine and examined the tetraploid cells produced, he could detect the level of SCEs in two different cell cycles. These frequencies turned out to be the same. A problem in interpreting these experiments was that, in 1972, Gibson and Prescott discovered that tritium incorporation evidently increased the frequency of SCEs.

More recently, a number of staining techniques have been developed that allow the investigator to distinguish between chromosomes with BU substituted for T in one strand of DNA and chromosomes with BU substituted for T in both strands. Zakharov and Egolina showed in 1972 that Giemsa stained the two chromatids differently with BudR (deoxyribonucleoside with BU substituted for T) was substituted for TdR for two rounds of DNA replication. Giemsa stained the sister chromatid with BU in both strands darkly but the sister chromatid with BU in only one strand lightly. Latt subsequently showed that the fluorescent dye, Hoechst 33258, stained the chromosomes very brightly when T was present, less brightly when T was in one strand and BU in the other, and barely perceptibly if both strands contained BU. Other techniques similar to one or both of these approaches have been worked out subsequently.

The studies of SCEs have given rise to a number of interesting observations. The SCEs have a very strong tendency to occur in junctions between euchromatin and heterochromatin, at least in muntjac and in kangaroo rat. These organisms have certain distinctive distributions of euchromatin and heterochromatin that make it possible to distinguish locations of these chromatin categories in chromosomes stained to reveal SCEs. Another observation has been that there are more SCEs than normal in a genetic disease, Bloom's syndrome, in which cancer incidence is high and there are many chromosomal aberrations. This result does not necessarily mean that SCEs and chromosomal aberrations have a common basis, however, since two other genetic diseases with the same symptoms (Fanconi's anemia and ataxia telangiectasia) lack SCE increases.

Even when no differences in basal levels of SCEs are found, it is sometimes possible to see differences between cells where the DNA has been damaged. Mitomycin C, a bifunctional alkylating agent that crosslinks DNA, causes fewer SCEs in young tissue culture cells than in aged ones, for example, although the basal SCE levels for the two types of cells are similar. Cells of patients with Fanconi's anemia are less sensitive than normal to Mitomycin C as an SCE inducer, but are more likely than normal cells to have chromatid breaks where an SCE might have been expected. With regard to chemically induced increases in SCEs, it has been proposed that mutagenic carcinogens can be detected by in creases in SCEs.

Concluding Remark

Cytological genetics is concerned with cell structures related to DNA replication, nuclear division, and gene expression. Mitotic chromosomes or meiotic chromosomes can be stained to reveal bands. Depending on the stain and the pretreatment, different regions of each chromosomes are stained. Characteristic band patterns can be used to identify particular chromosomes. Currently, C-bands are thought to be constitutive heterochromatin, G-bands are thought to be intercalary heterochromatin, and R-bands are thought to be euchromatin.

The polytene chromosomes generally, but not always, have visible puffs at the sites of RNA synthesis. The transcribed regions contain DNA that hybridizes with the RNA produced; other bands which are not transcribed also can hybridize with the product RNA. Transcription involves entire puffs, but much of the RNA product may be quickly degraded so that hybridization occurs only at one portion of the puff. The location of the mRNA-specific RNA polymerase can be shown via fluorescent antibody binding; it is found in puffs but also in bands that are not puffed.

Lampbrush chromosomes of meiotic I prophase in oocytes are in the process of transcription of their loops. Both mRNAs and rRNAs are synthesized. RNA transcription proceeds around each loop, starting at one end of the loop and terminating at the other end.

B chromosomes are supernumerary chromosomes, having little or no homology with necessary A chromosomes. In corn, the B chromosome is highly heterochromatic. It tends to enhance recombination in certain regions of the A chromosomes.

Aneuploids have either lost or gained one or more chromosomes from the normal set. Non-disjunction is a major cause of aneuploidy. Down's syndrome and other autosomal aneuploidies have far-reaching effects on the phenotypes of the affected persons. Sex chromosomes aneuploidies are also found in viable persons.

Chromosomal rearrangements can be detected cytologically. The human cri du-chat syndrome results from a short deletion, for example. Translocations and inversions can be detected via examination of synapsed homologs.

Mitotic and meiotic chromosomes have a fine structure consisting of microconvules, which are condensed loops. The loops are attached at each end to a scaffold of non-histone proteins; these loops are made of aggregated nucleosomes of the 250- to 300-Å fiber.

Meiosis consists of a long prophase I, subdivided into leptotene, zygotene, pachytene, diplotene, and diakinesis phases. Synapsis via synaptinemal complex occurs in zygotene; crossing over can be seen at pachytene. This prophase is followed by metaphase I; anaphase I and telophase I separate the homologs into different nuclei. Prophase II is often absent; metaphase II ensues, then anaphase II, and telophase II separate the sister chromatids.

DNA replication and transcription can be studied at the cytological level, via light microscopical autoradiography or via electron microscopy. Results are consistent with unineme chromosome structure.

Transcription requires rearrangement of nucleosomal beads into a smoother structure, but transcription need not begin as soon as this rearrangement occurs. Evidently chromosomes are divided into transcription units (blocks of adjacent DNA that are turned on or off simultaneously). Electron microscopical studies have shown that DNA replication and transcription can occur simultaneously.

Sister chromatid exchange (SCE) is more common than it was once believed to be, and can be detected via differential labeling or staining of the sister chromatids. Abnormally high levels of SCE are associated with some genetic diseases, notably Bloom's syndrome.

INDEX